Sex

and

Scientific Philosophy

Dr Giora Ram

IMEXCO General
Publishing
Israel

Books by Dr Giora Ram

ADHD - Children of Tomorrow
Published in Israel by 'Gvanim' - [2010]
http://adhd.imexco.com - DanaCode 0-00860000644-6

The House on the Hill
Published in Israel by IMEXCO General Ltd - [2010]
Poems and Love Letters
http://love-u.imexco.com - DanaCode 0-08000250081-4

My Love, My Wife, My Divorcee
Published in Israel by IMEXCO General Ltd - [2010]
Dating and Mating
http://my-love.imexco.com - DanaCode 0-08000250082-1

The Hungarian Connection
Published in Israel by IMEXCO General Ltd - [2010]
An Autobiographical Novel
http://huncon.imexco.com - ISBN 978-965-91623-0-7 (Amazon)

Copyright © 2012 by the Author

First published in Israel by IMEXCO General Ltd.

All rights reserved. No part of this publication may be reproduced, stored in a retrieval system, or transmitted, in any form or by any means, electronic, mechanical, photocopying, recording, or otherwise without the prior permission of the copyright holder.

Printed in Israel

ISBN 978-965-91623-1-4
http://philosex.imexco.com

Contents

1. Prologue — 5
2. Genesis — 7
3. Love and Sex — 11
4. Why Women Live Longer than Men — 20
5. Time, Space and Relativity — 24
6. I am Body and Soul — 31
7. Children of Tomorrow — 46
8. The Philosophy of Negotiation — 71
9. Scientific and Philosophical Insights — 84
   - 9.1 Spiritual evolutions and the power of belief — 84
   - 9.2 Time relativity of evolution — 90
   - 9.3 Islamic invasion — 92
   - 9.4 Diversification of democracy — 94
   - 9.5 The desire to live — 96
   - 9.6 Virus of the mind — 98
   - 9.7 The polarity of everything — 103
   - 9.8 Life-logging — 104
   - 9.9 Experience transfer to the next generation — 104
   - 9.10 Neurophilosophy — 106
   - 9.11 Brain-to-brain communication — 110
10. Chronic Pain Detection — 113
11. Epilogue — 118

*You don't need
a white beard or to be dressed in a white robe,
contemplating the meaning of life,
to be a philosopher* (author).

1. Prologue

Sex is the glue for the existence of humanity (author).

The word *sex* usually refers to the obvious meaning of biological reproduction of both humans and animals. In its wider context, however, it has penetrated almost everything in life.

We can read and hear sexual words and their diverse multimedia implications in books, theatres, films, advertisements and our own everyday language.

Our worldwide media are saturated with sexual imagery, from delicate erotic hints to detailed visual and printed pornography.

The word *philosophy* comes from the Greek *philosophia* (φιλοσοφία), which means 'love of wisdom'. It was used by ancient Greek philosophers to describe their way of life.

Philosophy is the study of interdisciplinary, general and basic human problems, asking questions about knowledge, values, reason and existence.

Certain philosophical and other insights in this book derive from extensive research by the author.

The scientific elements owe much to his university career and the studies and projects implemented in international industrial and other ventures described in his book *The Hungarian Connection* [2010].

Some of those activities were carried out over the last three decades at the Weizmann Institute of Science at Rehovot, Israel, the Royal Postgraduate Medical School, London, the European Space Research Organization, Darmstadt, and the Keck Center for Collaborative Neuroscience at Rutgers University, New Jersey.

The philosophical insights relating to inter-personal relationships and about mating and dating are based on the actual personal experience of the author over a decade. Part of this research and some actual cases appeared in his book in Hebrew entitled *My Love, My Wife, My Divorcee*, published in Israel [2010].

The chapter about the philosophy of negotiation is based on the author's negotiation skills and experiences in his international business endeavours. Particular experiences, especially regarding antique markets, are included.

In Tel Aviv, the author indulged his hobby by attending the local antique market every Friday, not only as a buyer but also as a seller. The monetary reward was insignificant and often did not cover the expenses and fees involved in participating in the market as a regular dealer but he very much enjoyed the atmosphere and his encounters with a diversity of people.

The chapter entitled 'Children of Tomorrow' is based on extensive research on ADHD and its variations. It includes a methodological non-drug-based treatment implemented successfully on the author's son. Part of this chapter is based on his book in Hebrew entitled *ADHD-Children of Tomorrow*, published in Israel [2010].

Dr Giora Ram, 2012

2. Genesis

2. בְּרֵאשִׁית

"וַיֹּאמֶר יְהֹוָה אֱלֹהִים, לֹא-טוֹב הֱיוֹת הָאָדָם לְבַדּוֹ; אֶעֱשֶׂה-לּוֹ עֵזֶר, כְּנֶגְדּוֹ. וַיִּצֶר יְהֹוָה אֱלֹהִים מִן-הָאֲדָמָה, כָּל-חַיַּת הַשָּׂדֶה וְאֵת כָּל-עוֹף הַשָּׁמַיִם, וַיָּבֵא אֶל-הָאָדָם, לִרְאוֹת מַה-יִּקְרָא-לוֹ; וְכֹל אֲשֶׁר יִקְרָא-לוֹ הָאָדָם נֶפֶשׁ חַיָּה, הוּא שְׁמוֹ. וַיִּקְרָא הָאָדָם שֵׁמוֹת, לְכָל-הַבְּהֵמָה וּלְעוֹף הַשָּׁמַיִם, וּלְכֹל, חַיַּת הַשָּׂדֶה; וּלְאָדָם, לֹא-מָצָא עֵזֶר כְּנֶגְדּוֹ. וַיַּפֵּל יְהֹוָה אֱלֹהִים תַּרְדֵּמָה עַל-הָאָדָם, וַיִּישָׁן; וַיִּקַּח, אַחַת מִצַּלְעֹתָיו, וַיִּסְגֹּר בָּשָׂר, תַּחְתֶּנָּה. וַיִּבֶן יְהֹוָה אֱלֹהִים אֶת-הַצֵּלָע אֲשֶׁר-לָקַח מִן-הָאָדָם, לְאִשָּׁה; וַיְבִאֶהָ, אֶל-הָאָדָם. וַיֹּאמֶר, הָאָדָם, זֹאת הַפַּעַם עֶצֶם מֵעֲצָמַי, וּבָשָׂר מִבְּשָׂרִי; לְזֹאת יִקָּרֵא אִשָּׁה, כִּי מֵאִישׁ לֻקְחָה-זֹּאת. עַל-כֵּן, יַעֲזָב-אִישׁ, אֶת-אָבִיו, וְאֶת-אִמּוֹ; וְדָבַק בְּאִשְׁתּוֹ, וְהָיוּ לְבָשָׂר אֶחָד." (בראשית ב יח-כד)

And the LORD God said, It is not good that the man should be alone; I will make him a help meet for him. And out of the ground the LORD God formed every beast of the field and every fowl of the air; and brought them unto Adam to see what he would call them: and whatsoever Adam called every living creature that was the name thereof.

And Adam gave names to all cattle, and to the fowl of the air and to every beast of the field; but for Adam there was not found a help meet for him.

And the LORD God caused a deep sleep to fall upon Adam, and he slept: and he took one of his ribs, and closed up the flesh instead thereof; And the rib, which the LORD God had taken from man, made he a woman, and brought her unto the man.

And Adam said, This is now bone of my bones, and flesh of my flesh: she shall be called Woman, because she was taken out of Man. Therefore shall a man leave his father and his mother, and shall cleave unto his wife: and they shall be one flesh.
(Genesis 2: 18-24).

This is what scripture says…but actually, it was not like this…

In a recent archaeological excavation of the author's feverish brain a secret scroll was found within certain synapses.

Here is a scoop!

For the first time since creation, we can reveal what really happened in Paradise. Here is the exact dialogue between God and Adam:

Adam: God I am bored, I have nothing to do, no electricity, radio and TV, no sport channels…only your divine silence is all around, all my testosterone is wasted, please do something…

God: What exactly do you want?

Adam: I don't know…create some divine creature, but not like me, maybe an interesting creature that can cope and struggle with my ego, one who'll help me to spend my spare time and will give me simultaneously intellectual stimulation, but not too much…

God: OK, I have an idea, but it will cost you three ribs.

Adam: What…? Three ribs…? That's too expensive.

God: I understand that you care about your ribs, but you have to learn that there is no such thing as a free lunch. If you prefer, I have something simpler for two ribs.

Adam: No, I am willing to give only one rib, but only under heavy sedation and if you guarantee no pain.

God: Agreed…

Adam opens his eyes and a gorgeous woman appears in front of him. One that wouldn't be accepted for today's catwalk, because she had more than satisfactory measurements in all the right places, far removed from the typically anorexic features so beloved of top modelling agencies.

His blood pressure is skyrocketing and with it his physiology…and he says to himself: 'What an idiot I am…if that's what I get for one rib… what I could get for three?…'

Ever since that day men have carried in their genes the same frustration. Although most of them find their own amazing Eve, they still seek other Eves without appreciating the first.

Ladies! The smirk on your faces is obvious… but you have no reason to celebrate. Men sometimes leave their spouses for good reason; every coin has two sides.

Only you can be blamed for this inter-relationship mess. If your mother hadn't tasted that damned apple…we would be still in paradise...naked…and it's no use blaming the snake…

Here is another version of the creation…Eve was created first. After a while, albeit having a great time in the Garden of Eden with the apple trees and the snake, she felt lonely. She turned and asked God to resolve her loneliness…
God: I'll create a man for you.
Eve: OK, will I get along with this man?
God: Yes, but you'll have to take good care of him. He'll do good to you, but he'll have certain qualities that you'll have to be able to handle…for example he'll have an inflated ego, he'll not be always gentle with you, sometimes he'll break your heart, but don't worry, I'll give you very simple operating instructions...
Eve: I don't know…I want but at the same time I don't want…is there anything else I need to know or do?
God: Because he has an inflated ego, you'll have to tell him that he was created first…
I am doing it just for you as woman to woman…

God created man in His own image, in the image of God He created him; male and female He created them (Genesis 1:27).

As amateur anthropologists, reading between the lines, we may conclude that man was initially created as androgynous (hermaphrodite). This assumption is supported by Greek mythology.

Hermaphroditus was the son of Hermes and Aphrodite or Venus, the love goddess. The water nymph Salmacis fell in love with him. For some reason Hermaphroditus wasn't too keen on her aggressive courtship, but she didn't give up; she attached herself to him and prayed to the Gods. Her prayers were accepted and their bodies were joined to form a bisexual creature.

According to this interpretation, the man created in Genesis was originally a bisexual creature with both male and female organs and only later was the female part separated to become an individual creature.

As the separation was not completely genetically clean, certain female features and qualities were left with the man and certain male features were left with the woman.

During the evolutionary process a wide spectrum of male and female features was created and diverse gene distribution took place among the sexes.

Because all humans are created equal and they are God's creations, men, women, gays, lesbians and all in between, we have to accept them in love, as they are our brothers and sisters, flesh of our flesh and blood of our blood.

We all come from the same place and when the time comes we will all return to it.

After conducting a few dialogues, monologues in some cases, with orthodox and other religious people, I found they expressed surprising empathy, understanding and consideration in terms of the diversity of the sexes. However, most religious people of most religions especially the radical ones are rejecting the notion of diversity of sexes and will not tolerate and accept the mere existence. In the complex relationship between men and women, there have been substantial changes since creation, some positive and some negative.

3. Love and Sex

Love was, love is and love will prevail (author).

The word *sex* usually refers to the obvious meaning of biological reproduction of both humans and animals. In its wider context, however, it has penetrated almost everything in life.

We can read and hear sexual words and their diverse multimedia implications in books, theatres, films, advertisements and our own everyday language.

Our worldwide media are saturated with sexual imagery, from delicate erotic hints to detailed visual and printed pornography.

Since Adam and Eve, the evolution in human sexuality and its by-products has superseded evolution in all other areas.

In general, there is no human love without sex; however, there is sex without love.

Much has been written about love and there is no need to repeat here the words of the world's great philosophers, lovers and poets.

It could be quite disappointing to analyse falling in love from the scientific perspective. If you were told that it was all to do with a very small area of the brain and a transient chemical process, it could well kill desire and the longing for love.

The psychologist Robert Sternberg defined love in the context of an interpersonal relationship as a combination of three elements: *intimacy*, *commitment* and *passion*.

Baruch de Spinoza, the seventeenth-century philosopher, identified three different elements defining love: *desire, joy* and *sorrow*.

In the love triangle, passion or desire plays a major role. Without passion, there is no love.

Eros or Cupid and his partner Psyche typify sexual love in Greek and Roman mythology. A unique version of love is the concept of platonic love, a relationship without sexual elements.

Aristophanes, the famous Greek comedy writer, told an interesting myth about the story of creation. Humans were connected in pairs, so each creature had two heads, four legs, four hands and two hearts. They were punished because they sinned against the Gods. Zeus separated them and since then each half soul has been searching for his/her mate. Therefore, true love is possible only between soul mates.

Love is me, love is you

There are no conditions in love, no *if, but* or *when*. Sometimes love is painful, disappointing and makes us lose our head.

Love is a function of age. Romeo and Juliet's teenage love is not the same as adult love. The attitude to the word *love* and its significance is different not only in certain age groups, but also in different cultures.

We tend to associate the word *love* not only with humans but also with animals, items, food, behaviour and films. Obviously, there is parental-family love, love of friends, love of neighbours and of course love of God.

Thou shalt love thy neighbour as thyself (Leviticus 19:18).

If, however, *thyself* is narcissistic, it might be a destructive love. On the other hand, if it is a healthy narcissism, it might be a positive and interesting type of love.

There are many types and groups of narcissistic behaviour defined and classified according to the amount of self-love involved.

According to Freud, our ego develops during infancy and therefore we do not possess the knowledge of self-love at birth. Later in life, our parents, friends and social environment mould our image of the aspired self, which builds and generates our ego.

There is nothing wrong with mild narcissism or mild ego. An egoistic couple can be quite happy. Even Freud realized that exclusive self-love might not be as abnormal as previously thought.

Narcissus was a handsome Greek youth, according to the myth, who fell in love with his reflection in a pool when he saw it for the first time. He died when he realized that he was unable to consummate his love. The narcissus flower bears his name.

Can we make ourselves fall in love with a spouse whom we only initially like?

The answer is yes. It may be done according to the principle of 'Fake it till you make it'. This may sound illogical or impossible to implement, but after self and mutual training you will see that in most cases it can work.

You start with small steps of positive and constructive communication with your spouse. Exchange mutual compliments and see the cup as half full. Think whether you can live without your spouse; convince yourself that you are lucky to be together. Create romantic situations; buy small gifts and show care and empathy. Invest in the relationship and say the words 'I love you'. Show continuous enthusiasm for the relationship, and be generous in showing affection. Write poems and love letters.

Show empathy and understanding regarding every action, saying or behaviour of your spouse; see the positive side even if you disagree. The more you persist and the more positive your attitude, the more you will drag your partner into this process.

You will march together along the path of love, each supporting the other with mutual consideration and encouragement.

Love brings more love

There is yet another possibility of finding and even creating love. This option is not so palatable and is quite difficult to exercise. If you succeed, however, you will attain a level of love that very few experience.

This kind of love is not only unique but long-lasting as well. It is ineffective to try it just once; it must be exercised often and practice will make it perfect.

How and what, you may well ask. The answer is found in the Four Seasons' song, 'Silence is Golden'. This is a type of love that is created from mutual understanding, consideration and respect. It is a unique love that evolves in a process of mutual *silence*.

Love is also loyalty; obviously, love will be encouraged and strengthened if loyalty is mutual. Love is about loyalty and support which are unconditional and unlimited. It is the kind of love that is not subject to any preconditions.

This blind and total idealization of loyalty we can see in films, especially the black and white films of the fifties. Women were depicted as supporting everything their men said or did, even if they disagreed. They 'stood by' their men.

Most importantly, those women would never air disagreements in public or in front of strangers. They did however express their dissatisfaction gently in the privacy of their home. Insulting your spouse in public, in front of friends or strangers, betraying trust and loyalty or hurting your spouse's self-image and ego, are a sure recipe for divorce.

How do we know if a man really loves a woman?
When she wishes to leave him and he agrees...

The following is about loyalty and love:

The Station

*To my own beloved wife
I, your lover, am dictating this letter as
I was wounded during the last battle
And a terrible thing happened to me*

*Both my hands were cut off
And my left leg is being amputated
Tell me, my love, if you will still accept me
Or if I should stay here forever*

*The woman without hesitation writes
Stay there and never come back
I am still young and love travel and
dancing and poetry*

*On a bright day at high noon
While the sun stood high in the sky
A train arrived at a station and
A war hero climbed down quickly*

*He walked without stick or support
In his hand, he held his case
The letter he wrote home
was sent to test his wife's loyalty...*

There have been many famous and unique loves since the creation of the world. Among the Biblical examples, we are familiar with the loves and desires of King David, the story of Bathsheba and the love of Jonathan...We know what happened to the love of Samson and Delilah, and about the amorous adventures of Moses.

We read about our ancestors, their loves and desires.

The love stories of Shakespeare's Romeo and Juliet and Nabokov's Lolita fall into this category of love, but offer a sensual and hormonal perspective.

It is not true that people stop pursuing dreams because they grow old, they grow old because they stop pursuing dreams–
(Gabriel Garcia Márquez - 1982 Nobel Prize for literature).

Another version of the above would be:

It is not true that people stop falling in love because they grow old, they grow old because they stop falling in love.

The ability to love is given to all, but not everybody knows how to recognize true love.

You need two to tango

Being in love is the result of two souls reaching a high level of successful human communication. One side is the transmitter and the other side is the receiver and vice versa.

Successful communication is achieved only if the one who is giving love is able to transmit it properly through the right channel and at the same time the one receiving love has their receiver turned on to receive the love message, interpret it and respond accordingly.

We all have problems at both the transmitter and the receiver ends. Sometimes the transmitter is inoperative and sometimes the receiver is turned off.

Much stimulation is needed to generate the right *click* that we need to get into the proper state of mind, to be able to love and be loved, as in the famous lyrics of 'Nature Boy' (song by Eden Ahbes, published in 1947):

The greatest thing you will ever learn,
Is just to love, and be loved in return.

This stimulation may be located in diverse areas of visual appearance, smell, physical and mental attraction, race and origin, family involvement and there may be many other logical and illogical reasons why that unique *click* which generates the essential ingredients of falling in love is triggered.

Love is… There is a long list of adjectives following the definition or explanation of what *Love is*. Love is… me, you, us, great, strong, healing, happiness, madness, love is everything.

Love is friendship, respect, communication, admiration, together, love is missing you. Love may decay with time or distance, but the right trigger may ignite the fire again. True love cannot be stopped.

Love is all we need

Love is age-dependent, takes different forms and is subject to different interpretations. A baby's love of a feeding and caring mother is different from hormonal teenagers' love, which is not the same as mature love or love in old age. Nor is male love the same as female love. Each may say to the other 'I love you' but mean different things.

True love combines body and soul. Great sex is when two souls are interlaced as well as bodies. Sexual desire may decay with time. Sometimes the cause is physiological, but usually it is psychological. The body wants sex but the brain is disconnected. In such cases the usual advice is to attempt the act, in spite of low desire or no desire at all. The more you try, the more the desire will come. There is nothing wrong with faking; the experience will be real with time. The worst thing is to accept the numbness and accept the lack of desire, which leads eventually to separation.

Women are from Venus (♀) and men are from Mars (♂)

Research shows that females use the left side of the brain more often than the right; therefore they perform better in the linguistic area. Males on the other hand use their right hemisphere more and are more competitive and aggressive and responsible for most wars.

Female memory is different from male and many experiments have proved that. Put a male and female into a room with people and objects for five minutes and ask them when they leave to report what they saw and heard. The amount of detailed information that the female will produce will be about twice as much as the male's.

Obviously, the male will notice the females and their location in the room, whereas the female will notice both sexes. In addition, she will also remember colours, hairstyles, and conversations.

There is a saying about the female memory:

She never forgets and never forgives

She will remember what you told her about a decade ago; everything is saved in her brain and will be retrieved quickly at need. Women are like angels; they multi-task and are able to perform complex tasks. Women are more spontaneous than men, who are more focused, persistent, sometimes stubborn and very target- and goal-oriented.

Men are focused on conquest whether it be a work goal or a woman. This feature has led to the assumption that the female brain has more connections between the two hemispheres than does the male brain.

This quality is extremely important, especially in case of brain injury. Those extra connections may enable the opposite side of the injury to assume the tasks of the injured region.

There are some lucky couples who, after years of being together, reach such a level of togetherness that they can read each other's mind.

Most of us can only envy those couples whose love and understanding has enabled them to develop some kind of telepathy.

The rest of you, the non-telepathic, must learn the inter-galactic telepathic language spoken between the sexes from Mars and Venus in order to succeed in finding the right partner and maintaining the relationship for *as long as you shall live.*

Love is a rare and precious thing. If you are among the lucky ones, you must do everything, but everything, to preserve it.

*What cannot be achieved with love
can be achieved with more love* (author).

4. Why Women Live Longer Than Men

It is an acknowledged fact that women live longer than men. The reasons are manifold. An attempt is made here to describe briefly some of the factors identified in published studies and to suggest new directions for further research.

It is believed that those studies may lead to the discovery of new factors explaining the discrepancy in lifespan between women and men. One such is the area of foetal *microchimerism*, which is the persistence of certain numbers of foetal cells in the mother after pregnancy. Another may be found in further studies about the system involved in enhancing a mother's immune system, during and after pregnancy.

Further studies may find that women who give birth live longer than those who do not. In addition, those women who live to be over 100 may have special genes that slow down the ageing process and reduce the probability of heart disease, stroke, cancer and Alzheimer's disease.

The X chromosome may be another influential factor in female lifespan. Because women have two X chromosomes, if there is an abnormality on one of them, the normal one can be used rather than the faulty one. In this case the woman becomes a carrier of the disease but not its victim.

Serious studies point toward the *menopause* as a major determinant of lifespans. Very few species menstruate apart from women and certain whales.

The evolutionary force needed to pass on genes and the need to stay alive and bear the maximum number of children may be some of the natural forces that enable women to live longer than men.

In the USA and other developed countries, average life expectancy is about eighty years for women and about seventy years for men. In Sweden, however, during the eighteenth century it was about thirty-seven years for women and thirty-four for men.

The probability of women outliving men during the first twenty-five years is more than four times higher, mainly because of men's *testosterone*. This factor decreases with age, but the gap still remains in favour of women. The sex hormones are a clear factor to consider; young males are aggressive, increased levels of harmful cholesterol leading to heart disease or stroke.

The female hormone *oestrogen* lowers harmful cholesterol and raises 'good' cholesterol, however. Recent studies suggest that oestrogen treatment after menopause might reduce the risk of death, especially from heart disease or stroke.

Females have hearts that last longer and better blood vessels, possibly owing to the oestrogen hormone. Even diseases such as heart disease, stroke, cancer and diabetes ultimately kill more men than women.

The advantage women have over men does not apply, however, to women who smoke, drink, are overweight or under excessive stress, typical of women who work in domains previously dominated by men. Actually, there is no gap at all among male and female smokers.

There are however certain countries where there is almost no such gap, such as India and Pakistan, where sexual discrimination and practices such as female infanticide and bride-burning are rife.

Mortality rates may also be affected by chromosomal differences between men and women. They both carry genetic mutations that may cause certain life-threatening diseases.

Females have two X chromosomes, so if there is an abnormality in one of them the normal can be used, making the woman a carrier only of the disease. Men have one X and one Y chromosome, so they cannot use an alternative chromosome, if a gene is defective. The above-described differences in longevity have been observed in most animals.

Another fact is that men are much more likely than women to engage in risky and violent behaviour, consequently increasing the male death rate. More men than women also die in car accidents, homicides, and even suicides.

Foetal microchimerism is the persistence of certain numbers of foetal cells in a mother after pregnancy. A number of recent studies suggest that it may be the cause of some autoimmune diseases. Long-term persistence of foetal cells in healthy women is a contraindication, however. The long-term persistence of foetal cells may also have significance in developing the tolerance of the foetus.

If microchimerism is owed to the transfer of cells between mother and foetus, further studies might indicate the effect of those foetal cells in enhancing the mother's immune system during and after pregnancy. We know that during pregnancy cells can migrate between mother and foetus in *both* directions.

As pregnancy advances, the cell transfer rate from foetus to mother increases. In most cases, the foetal cells are compatible with the mother's immune system, so the mother's body does not reject them.

It is believed that such women, especially those who live to over ninety or a hundred, may have special genes that slow down ageing and reduce the probability of heart, stroke, cancer or Alzheimer's disease.

Further studies might show that women who give birth will live longer than those who do not. It might also be that those who give birth at least to one male will live even longer.

The detection of Y chromosomes or male DNA in women *after* pregnancy, even in a woman who had her last son many years prior to blood sampling, may support this theory.

In general, stem cells can be replicated and they are capable of long-term self-renewal. As they do not have any specific structures, they can evolve to specialized cells such as heart muscle, nerves or blood cells.

The sources of blood stem cell transplants are bone marrow, peripheral blood and the umbilical cord of newborn babies. A new technology that has generated new businesses is the collection of umbilical cord blood stem cells from babies, which are stored for future use.

The advantages of stem cells from the umbilical cord in conjunction with the fact that there is a transfer of cells between mother and foetus may be another factor explaining and supporting the theory that women who give birth will live longer than those who do not.

Finally, it should be emphasized that the reason that women live longer is based on many factors, only some of them mentioned above.

The Bible (English version) says (Genesis 16): '*Unto the woman He said: I will greatly multiply thy sorrow and thy conception; in sorrow thou shall bring forth* **children;** *and thy desire shall be to thy husband, and he shall rule over thee…*'.

The Hebrew version has: '*… in sorrow thou shall bring forth* **Boys**'.

Although some may claim that 'Boys' or *Banim* is the generic name for children, one may wonder if the original was indeed 'Boys'…

The longer lifespan of females might be part of some grand Darwinian scheme whereby mother nature is rewarding motherhood, and especially those women who give birth to at least one male; mothers may gain another five years of life for their effort…and they well deserve it…

5. Time, Space and Relativity

This chapter offers some thoughts and observations about time and time-related phenomena. It attempts to address certain questions on the continuity and relativity of time before a time-related problem-solving model is presented.

We use the word *time* directly and indirectly very often in our daily conversation and throughout our lifetime: time is money, time of life, time after time, between times, gain/lose time, good/bad time, slow/fast time, right/wrong time, before/after time, present time, past time, real time, on time, in no time, kill time, any time, every time, plenty of time, timeless, time limit, time cycle, time cures and time flies...

Time is depicted by artists in various ways, among them the famous 'melting clocks' by Dali. We can distinguish between pure time, relative time and absolute time.

Time measurement is the unit of time to which all time measuring devices ultimate refer.

Time is a point at or a period in which things happen, a repeated instance of anything or a reference to repetition, the state of things at any period.

Space is that part of the boundless four-dimensional continuum in which matter is physically rather than temporally extended.

Relativity recognizes the impossibility of determining absolute motion and leads to the concept of a four-dimensional space-time continuum.

The special theory of relativity, which is limited to the description of events as they appear to observers in a state of uniform motion relative to one another, is developed from two axioms:

1. The law of natural phenomena is the same for all observers.
2. The velocity of light is the same for all observers irrespective of their own velocity.

Space and *time* in the modern view are welded together in a four-dimensional space-time continuum. There is no clear distinction between a three-dimensional space and an independent time.

Time means different things to different 'observers'. This may not agree with the axioms (on which the special relativity theory is based) described earlier, at least not from a psycho-philosophical point of view. These 'observers' may include: people (humans), animals, plants, clocks and other beings outside our time universe. Time seems to be different for different people: age, education, origin, mental stage and religion may all have an effect.

Time appears 'slow' when we are young and 'fast' as we grow older. Time seems to be passing faster when we are enjoying ourselves or when we are busy, as opposed to when we are bored or idle. The description of time-related events in the history of humankind differs in different cultures.

For example, Adam, the first man, is described as 130 years old when his son Seth was born and Adam lived for 930 years! Seth lived for 912 years. The famous Methuselah lived for 969 years. Noah was 500 years of age when he had his three sons and he died at the age of 950.

Was life expectancy in those years really so long, or was the time measurement of a 'year' different?

Was earth moving in a different orbit?

These life expectancies are about ten times higher than ours today. Could it be that earth was then closer to the sun? Or moved in a faster orbit with different gravitational forces?

Was it a 'figure of speech' to indicate a long-lived person, or was the year count based on the moon and not on the sun?

Time is different for animals and plants as we can see from their lifecycles, behaviour and responses which are not what we might expect.

Clocks and other similar instruments measure time and tend to be almost identical in terms of information about it.

This is to be expected as we designed them all for the purpose of measuring time defined to be consistent within our universe.

If, however, earth's orbit changes, time will change as well. This will cause changes in our current concept and evaluation of distances expressed in light years, and could offer opportunities to reach far-off galaxies.

Time seems to be continuous, but is it?

We divided our earth year into subsets in different units: months, weeks, days, hours, minutes and seconds.

Scientists continued the division into one thousandths of a second or millisecond, a nanosecond, which is one thousand millionths of a second, and a Pico second, which is one million millionth of a second (one and 12 zeros).

Time is continuous with respect to our universe and within it, and it is relative to our observations.

When we observe a moving object between two points we 'see' it travelling all the distance between the two points, so we assume that this continuity of observation means that time is continuous.

This may not be the case, however, if we perform our observation in another galaxy or in another dimension, where these rules are not necessarily valid.

In the digital domain, as opposed to the analogue domain, we may observe the same continuity of moving objects. The time is digitized, however, and between two consecutive time points there is a gap of a certain fraction of a time unit, equivalent to the sampling resolution, where 'anything might happen'.

For other creatures these time gaps may represent their entire lifecycle, or we may be living within our time with another life form, whose time resolution fits with our 'dead times', which are our time gaps. Television is viewed as continuous moving pictures, whereas actually it comprises discrete individual pictures, projected at thirty frames (or more) or pictures per second.

Time can be measured, viewed and evaluated. The observer's tools for the evaluation of time are his/her senses. Unfortunately, senses can be fooled.

Strobe light projected onto a rotating disk will generate the illusion of a still disk. Are our other observations wrong or at least inaccurate, then, particularly if we are a small subpart or subspace of a much larger and more complex galaxy?

In the laboratory, we have successfully accelerated and slowed down certain processes, such as chemical or other natural processes. These experiments offered the possibility to control processes which were functions of time.

Certain processes were successfully reversed to what they were before, indicating 'pseudo going back in time', which is not really going back in time, but it looks like it.

The introduction of computers generated a revolution in time-related processes and enabled not only the observation of past and present time-related phenomena, but also predictive processes, which are future time-dependent scenarios.

Time affects our entire lifecycle, our birth, our life and death. Our heart beats almost once every second and our inner biological clock operates throughout our life. If we overturn this clock by flying to another time zone, our body suffers a phenomenon known as jet lag and it takes some time to adapt to its new condition. Time affects most of the processes and phenomena on earth, some faster and some slower. If there are time-independent phenomena or a phenomenon that until today has seemed to be unaffected by time, then these scenarios must be classified as 'past, present and probable future'.

As the observer's time is limited, we are unable to analyse these timeless phenomena without using assumptions and predictions. 'The sun was'. It may have been true since creation. 'The sun is – it exists'. It is true from our current observation. 'The sun will be'. Maybe yes and maybe no.

Statistics provide us with the probability, P, that there will be a sunrise tomorrow: it is 'almost definite' or it is a number close to one. This predictive calculation is based on the observation that the sun has risen n times until today, where n is a large number: $P = n/(n+1)$.

According to Einstein, time is more like a river, flowing around stars and galaxies, speeding up and slowing down as it passes massive bodies. One 'second' on the earth is *not* one second on Mars.

'I was, I am and I shall be' is written in the Bible. This infinite existence refers only to God.

All materials, including all known life forms and other mass owned celestial bodies are time-dependent.

Throughout our history we have asked ourselves many questions about many subjects relating to sciences, to the so-called supernatural, to creation, to travel through time and suchlike.

As humans we have tried throughout history to answer these questions using logical knowledge and experience. The fact is that most of these questions remain unanswered today.

When we approached religious people or scholars their answers depended largely on belief in God or the supernatural and were far from satisfactory.

The answers given to us, particularly during our childhood, were 'only God knows the answer' or 'believe it' and so on.

These responses only increased our curiosity and determination to make some sense of it all. The more educated and experienced we became, the more the conundrum seemed sophisticated and impossible to understand. It looked like the easiest option would be to stick with the religious domain and drop the subject.

Using some philosophical-mathematical tools, we will attempt to address some of the issues relating to solving open problems. An open problem is categorized as such if and only if it has not been solved by an expert or a professional individual and remains unsolved today.

By solving, we also mean proving that the specific problem is solvable if not today then at a later time when the necessary tools for its solution are available.

Philosophical problems may be presented by mathematical-logical models, whereby well-defined tools of measurement, quantification, proven theorems and axioms offer 'questions' and 'answers' or 'problems' and 'solutions'.

A *point* may be defined as a problem or a solution.

A *line* is defined as a group of *points* with common features.

A *plane* may include a line or lines, which have common features, so a *plane* is defined as a group of *points* or *lines* with common features.

We define a line here as representing a specific problem with no available solution.

The length of the line is proportional to the *time* spent from the moment the question became an open problem (finite) and the *time* required for the solution (assumed to be infinite until a solution is found). This line is finite at one end and infinite at the other. We define a line representing a specific solution to a problem similarly.

A match or a solution matching a problem will occur if and only if two appropriate pairs of lines (question and answer) out of the entire universe of lines intersect. When this happens the question is no longer unanswered and is removed from the group of unanswered questions and problems. The answer remains if and only if we assume a one-one correspondence between question and answers, namely that an answer is an answer only to a specific question; otherwise it also can be removed.

If there is no current answer to the question, we say that the two appropriate lines are *parallel*. Given the theorem that two parallel lines will intersect at infinity, this should imply that *all* questions will be answered *eventually*.

As our lifetime and those of our successors are finite, it is unlikely that we (humans) will reach the infinity point in terms of finding and understanding all the answers.

This naturally relates only to those answers that need infinite time to reach a solution (intersection). If, however, we assume that we *are* at the point of intersection then this would imply that we have eternal life...

From the above, and adopting a religious-philosophical approach, one could infer that there must be someone 'above' us in this universe, who *was* before all of us and who *will be* after us at the intersection-infinity point.

6. I am Body and Soul

Since time immemorial the philosophical question of the existence of the soul has troubled humanity and many philosophers and theologians have debated the subject. When you think about it, you realize that there must be *something* in addition to your physical body. The terminology is not important in this discussion and one may equate the word *soul* to *mind* or any other word suited to one's opinion or religion.

The relationship between mind and body and the existence of the soul are the subject of this chapter.

> *And the Almighty formed the man of dust from the ground, and He blew into his nostrils the **soul** of life* (Genesis 2:7)

According to this verse, the soul cannot die. The argument for that claim is that since God *blew* life into man, He blew it from Himself and since God is eternal and spiritual the soul cannot die as it is part of God. A soul is divine energy or a piece of God within us. King Solomon wrote:

> *The dust will return to the ground as it was, and the spirit will return to God who gave it* (Ecclesiastes 12:17).

Kabbalah in Hebrew means *receiving*. Kabbalah addresses the nature of the universe and the human beings who live in it.

There are many diverse aspects of life, all of which are addressed by Kabbalah. It tries to explain among others the relationship between God the Creator and His mortal creation.

A mystical concept within the Kabbalah of Judaism is the *tree of life*, which is used to explain the origins of God and the manner in which He created the world out of nothing. The tree presents ten *sephirot* (counting) representing cardinal attributes, showing their twenty-two paths or interrelationships. The Kabbalists developed this concept into a full model of reality, using the tree to depict a map of Creation (Fig. 1).

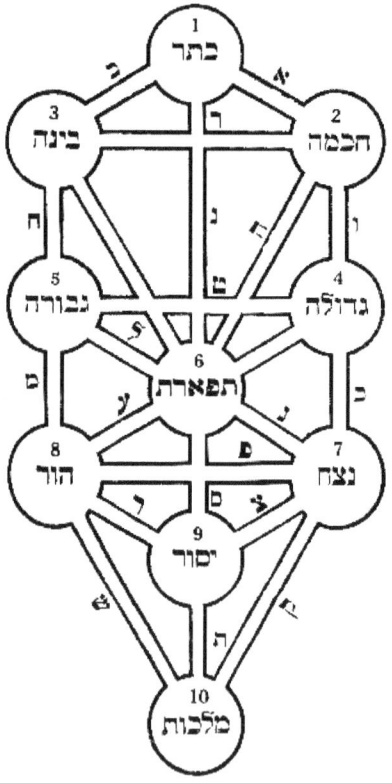

Figure 1. Kabbalistic map of Creation

The meanings of each sphere or attribute are:

1. Keter = Divine crown
2. Chochmah = Wisdom
3. Binah = Understanding and Daat = Knowledge
4. G'dula = Greatness and Chesed = Kindness
5. Gvurah = Strength and Din = Justice
6. Tif'eret = Beauty and Rachamim = Mercy
7. Netzach = Eternity
8. Hod = Glory
9. Yesod = Foundation
10. Malchuth = Kingdom, God's Presence.

The five levels or the five dimensions of the soul according to Kabbalah are:
Nefesh (soul), *Ruach* (spirit, wind), *Neshamah* (breath), *Chayah* (life) and *Yechidah* (singularity, unit).

Nefesh is the soul of physical life or instinct. *Ruach* is the emotional spiritual self. *Neshamah* is the intellectual self. *Chayah* is our rational self, desire and faith. *Yechidah* is our soul's unity with its source, and is part of God.

These five levels of soul can be described as ascending levels of awareness of and communion with God. When a child is born s/he is given a *nefesh* and expected to live in a physical and creative world or world of actions, which is the lowest level. His/her deeds will be taken into account when s/he is transferred to the next level.

Ruach is the next level of soul, which is on a higher plane than *nefesh*. The world of *ruach* is the world of *yetzirah* (production) and relates to morality. It is a stage which entails emotions and love, especially loving God with all one's heart and soul.

Neshamah is the level at which we contemplate the manifestation of divine energy in the world of *Beriyah* (creation).

Neshamah is the conceptual grasp of the intellect and can be described as a stage from which to commence communion with God.

Chaya is the level where the soul reaches the point where no ego, self or identity can exist. This stage is when one realizes the meaning and the truth of things as being in communion with God. The realization includes awareness of the divine energy and its unlimited capability in an endless universe.

Yechidah is the final stage of merging and being in communion with God. Realizing the infinity of light and purity, the essence of soul, is the fifth level.

According to Plato (c. 427-347 BC), the soul consists of three basic parts or energies: *reason*, *emotion*, and *appetite*. He saw the soul as the essence of a human being. When the body dies, the soul is reborn in a subsequent body.

The Platonic soul comprises three parts:

- The *logos* (mind, or reason)
- The *thymos* (emotion, or masculine)
- The *eros* (appetite, or feminine)

A peaceful soul must balance these three elements. The reason or the mind carries the greatest weight and it is expected to control the other two energies of emotion and appetite.

Plato claimed that as the body is material and the soul is an idea the latter must be immortal. He also claimed that the soul and the body are together only temporarily and they will be separated upon death.

The existence of the soul is independent of time and space and so it may realize and be exposed to the ultimate truth.

René Descartes, the seventeenth-century philosopher wrote about the *Passions of the Soul* in 1649.

The Nineteenth Article - Of the apprehension.

Our apprehensions also are of two sorts: the soul is the cause of some, the body of the other. Those whereof the soul is the cause are the apprehensions of our wills and all the imaginations or other thoughts thereon depending.

For we cannot will anything but we must at the same time perceive that we do will it. And although in respect of our soul it be an action to will anything, it may be said also a passion in her to apprehend that she wills.

Yet because this apprehension and this will are in effect but one, and the same thing, the denomination comes still from that which is most noble. Therefore it is not customary to call it a passion, but only an action.

The 34th Article – How the soul and the body act one against another.

Let us then conceive that the soul holds her principal seat in that little kernel in the midst of the brain, from whence she diffuses her beams into all the rest of the body by intercourse of the spirit, nerves, yea and the very blood, which participating the impressions of the spirits, may convey them through the arteries into all the members.

Let us here add, that the little kernel which is the chief seat of the soul hangs so between the cavities which contain these spirits, that it may be moved by them as many several fashions as there are sensible diversities in objects.

> *But withal, that it may be moved several ways by the soul too, which is of such a nature, that she receives as many various impressions (that is, has as many several apprehensions) as there come several motions into this kernel. As also on the other side, the machine of the body is so composed that this kernel being only divers ways moved by the soul, or by any other cause whatsoever, it drives the spirits that environ it towards the pores of the brains, which convey them by the nerves into the muscles by which means it causes them to move the members.*

Descartes was not the first to make the distinction between body and soul-mind, but his particular take on it was more influential on later philosophers. Combining or equating mind and soul is also debatable. The claim that the mind is the rational part of the soul is quite acceptable. Mind and body can co-exist in this life but exist apart from one another in another life or in the next life. This notion is partly supported by Kabbalah.

Discussing rational psychology, Immanuel Kant (1724-1804), identified the soul as the 'I' in a fundamental sense. He claimed that we cannot prove or disprove the immateriality of the soul.

The body is an empty shell and the soul energy is the intelligent energy of creation, which exists everywhere. Being in love and participating in mystical experiences allows us to feel the existence of our soul.

Energy is the communication or the bridge between our body-matter and our spirit-soul. Accordingly, when the soul leaves the body, the body dies, as it cannot exist without the soul. The body has needs, and sometimes it becomes sick.

There are many hopeless medical cases where the patient has had a miraculous recovery, because the soul energy came to help, as it was too soon for the patient to die.

The soul has needs as well. If one attends to those needs, in return the body is healthier. The converse is true as well.

A healthy soul requires a healthy body, as the Latin has it: *anima sana in corpore sano*, and in Hebrew: נֶפֶשׁ בְּרִיאָה בְּגוּף בָּרִיא

We know that mind and body are connected. Our health is strongly dependent on how we deal emotionally and physically with various situations and conditions. Our immune system is affected when we suffer continuous stress. General illnesses can be largely stress-related. Stress is not always negative, however. Stress is a predictable, normal and sometimes desirable human response. Stress is activated in various circumstances, such as feelings of danger or urgency. In such cases stress protects us by alerting us to danger and keeping us focused.

When stress is continuous and unrelenting, this is the point at which it starts to be harmful. It can have a negative effect in both physical and psychological terms. There is a link between stress and life expectancy. Most researchers believe that stress is a major factor in reducing life expectancy.

The author, however, believes that a certain amount of stress in reaction to certain activities can have the opposite effect. Both ends of the spectrum, namely, no stress at all and extensive stress, are negative factors which influence our way and quality of life. Some stress in a multi-tasking environment, however, can have a positive effect on our life expectancy.

Being active in multi-disciplinary areas not only gives us positive emotions and satisfaction but could also extend our life expectancy.

This assumption is partly supported by the fact that diverse activities mean we activate different brain areas.

Research has shown that negative emotions such as anger and unhappiness have a significant impact on our nervous and immune systems and are probably the cause of certain diseases. On the other hand, positive emotions such as happiness and harmony are elements which have a positive influence on our health.

One way to maintain good health is therefore to adopt a positive mental attitude. As the saying goes: 'Don't worry, be happy'.

When scientists speak of the soul outside the cultural, psychological or religious context, they generally treat it as a synonym for the mind. When we can decipher the brain's functionality, we should learn a great deal about the soul as well. Neuroscience research in the future might reveal the relationship between the soul and the mind and assist our understanding of the soul.

In neurobiology, the material functions of the mind could be a representation of certain mechanistic properties of the brain. The brain essentially shuts down the area which generates conscious thought during deep sleep and reactivates it on dreaming or waking.

The relationship of the brain with the mind is similar to that of computer hardware with software. This analogy of the mind as software is debated by scientists, who claim correctly that the human mind has powers beyond any artificial software.

Time and *matter* are the most fundamental concepts in physics. Quantum theory has produced many novel insights into these concepts in non-relativistic, relativistic and cosmological contexts.

Time is a subjective entity. Its purpose is to organize events and scenarios in a certain sequential or parallel order. In a different multi-dimensional universe, time may have different meaning, such as events that exist simultaneously.

Such concepts in terms of the theory of relativity can lead us to a hyperspace in a universe beyond space and time. Consequently, death can be viewed as merely a change of phase.

Writings about death are manifold but all are based on speculation and hypothesis. Nobody has died, been cremated or buried and come back to tell the story (I exclude here the Christian claim that Jesus Christ rose from the dead).

Those who compare death to a long sleep do so because we do not have anything else to compare it to.

Many adjectives and images surround the word *death*: *Death is not as many imagine it to be*; *Death is a temporary state*; *Death is a transition*; *Death is beautiful and eternal love*, etc. All these comforting descriptions make us, the living, feel good and less afraid or worried about the inevitable end of our physical body and our life as we know it.

A strange phenomenon has been reported by many individuals with regard to what is known as the *near death experience*.

This phenomenon has appeared in various forms, in different countries and cultures, worldwide. A near death experience (NDE) usually refers to a diverse range of personal individual experiences with multiple strange sensations: out of body experience, feelings of levitation and the presence of a light.

The mainstream scientific community regards such experiences as hallucinations, whereas paranormal specialists and some more maverick scientists would consider NDE as evidence of an afterlife.

Those who 'came back' when they were pronounced clinically dead were not really dead; possibly they were on the verge of death but were certainly not dead.

These 'dead' people report similar near death experiences. We can conclude from the similarity of those experiences certain things about our brain activity under stress and our self-awareness of approaching death.

There is an evolutionary connection between body and soul. People with a 'broken heart' will suffer physical pain. People with chronic pain will say that not only is their body aching but their soul as well.

Reincarnation and the existence of life after death are basic tenets of many religions.

In Judaism, Kabbalah describes the reincarnation of the soul as 'Gilgul Neshamot', whereby the soul of the dead person begins a new life in the physical world in a new body. Successive reincarnations allegedly eventually lead to the purification and liberation of the soul.

According to the Talmud (c. 500 C.E.) the soul is brought for judgment after death. The *good* souls may enter the next world and the *bad* souls need to be re-educated to enter the next world or the world to come. Some consider this period of re-education as a punishment for wrongdoing.

The concept of eternal damnation adopted by many religions is not widely accepted in Judaism.

Judaism believes that eternal punishment is reserved for a small group of villains and evil leaders, who have committed acts of pure evil against humanity. They will spend time in Hell (*Sheol*), which has four cells or compartments, for four different categories. The first is for saints who are waiting for resurrection in Paradise, the second is for the virtuous waiting to be rewarded, the third is for the wicked who await punishment, and the fourth is for the wicked who have already been punished. The last are those who will not be resurrected on Judgment Day.

According to Maimonides, a Jewish philosopher of the medieval period, when the soul is separated from the body it continues to live on in spiritual terms and is not physically resurrected. The Zohar (splendour or radiance in Hebrew) is a collection of written thoughts, commentary, and mystical views on the Torah. It serves as the basis for Kabbalah.

The Zohar describes Gehenna as the place of the soul's purification and not a place for punishment. Plato and other ancient Greek philosophers tried to prove the existence of reincarnation in philosophical terms.

According to Islam, the purpose of the creation of humankind is to worship Allah - the creator of heaven and earth.

Life on earth is a test of each individual before being sent to Jahannam or Jannah, which is Paradise in the afterlife.

In recent years, we may notice the 'evolution of the soul' and increase in the need for spiritual nourishment commensurate with the increase in life's difficulties. Gurus, rabbis, kabbalists and priests alike support this idea. Therapists and pseudo-therapists from diverse fields exploit those needs and weaknesses, usually for monetary gain.

The main purpose of mentioning the above views was to provide a background to the chapter's subject. It is difficult to argue against religion and belief but as almost all religions believe in the existence of the soul is it possible that all of them are wrong? And if they are right, what is the origin of their belief? Do they have any other sources apart from the Old and New Testaments?

Let us look for a different angle in the search for the existence of something in addition to the physical body.

The following table shows the constituents of a human body weighing 60 kg:

Constituent	**Weight**	**Atoms (%)**
Oxygen	38.8 kg	25.5 %
Carbon	10.9 kg	9.5 %
Hydrogen	6.0 kg	63.0 %
Nitrogen	1.9 kg	1.4 %
Other	2.4 kg	0.6 %

The human body is a living biological system. It can metabolize food and transform it into the energy needed by the body to function.

Its functions include pumping blood, transferring food to all body parts, maintaining body temperature, cleaning the body of undesirable substances, and protecting the body from external and internal injuries and diseases.

It can perform many more complicated functions, such as firing neurons in the brain, storing, processing and retrieving audio-visual and other information and, remarkably, carry and transfer genes to the next generation.

The Hindu religion requires cremation of the body as fire is helpful in transferring the soul to the other world. The Christian faith is not prescriptive, some branches discouraging or forbidding cremation and some permitting it.

The ancient Egyptians had complex and relatively unusual burial customs. They believed that certain procedures were essential to ensure immortality. Accordingly, they performed elaborate mummification and buried the dead with specific artefacts that they might need in the afterlife.

The Jewish faith, especially Orthodox Judaism which strictly follows the laws, forbids burning of the dead and any other form of burial except regular earth burial as they believe in the bodily resurrection.

Albert Einstein spoke about this many times:

I cannot conceive of a God who rewards and punishes his creatures, or has a will of the kind that we experience in ourselves.

Neither can I nor would I want to conceive of an individual that survives his physical death; let feeble souls, from fear or absurd egoism, cherish such thoughts.

I am satisfied with the mystery of the eternity of life and with the awareness and a glimpse of the marvellous structure of the existing world, together with the devoted striving to comprehend a portion, be it ever so tiny, of the Reason that manifests itself in nature.

In a letter of February 5th, 1921 Albert Einstein wrote:

> *The mystical trend of our time, which shows itself particularly in the rampant growth of the so-called Theosophy and Spiritualism, is for me no more than a symptom of weakness and confusion.*
>
> *Since our inner experiences consist of reproductions, and combinations of sensory impressions, the concept of a soul without a body seems to me to be empty and devoid of meaning.*

In a letter found in the Einstein Archives with Einstein's hand writing, the following sentence is written:

> *I do not believe in immortality of the individual, and I consider ethics to be an exclusively human concern with no superhuman authority behind it.*

This letter related to a question he was asked as to whether he had considered the relationship of his immortal soul with its creator, and whether he felt assurance of everlasting life with God after death.

In spite of his vigorous denial of the existence of the soul, did Einstein know something that he refused to publish?

The Relativity of Life and Death, perhaps?

This question arises because his famous formula of $E=mc^2$ (where E is energy, m is mass, and c is the speed of light) can be equated to the energy released as the result of cremating a human body.

Quasi-scientific support is often given as evidence that the soul exists. The soul is energy and energy can only be transformed, not disappear.

Therefore, the soul lives on after the death of the body. In cremation, the energy of the body is released as heat.

The ashes generated as the result of cremation of human adults weigh about 2.7 kg (six pounds) for males and 1.8 kg (four pounds) females, and represent about 3.7% of the total body weight.

Cremated remains contain dry calcium phosphates with minerals such as salts of sodium and potassium. In addition, certain gases are released during the process, such as sulphur and carbon, which turn to oxidized gases.

The following experiment was conducted. A dying man, who had left his body for scientific research and believed that his soul would be released after death, was anaesthetized and cremated.

The cremation was held in a completely sealed environment with multiple sensors, including a high-resolution digital video camera. For comparison, two other dead bodies were cremated as well, one by one.

The ashes generated as the result of the cremation were exactly as predicted from the body mass. The heat energy released was measured as well.

The result was amazing!

When all the parameters were calibrated and calculated, the result showed clearly that there was an outburst of unexplained energy when the dying person was cremated.

The equivalent energy released by the two dead bodies was as predicted.

It was obvious that the cremation of the dying person generated some unexplained extra energy.

Was it his soul?

Note: This experiment was performed at a secret location in the author's feverish brain, deep in his synapses.

7. Children of Tomorrow

Autism, ADD, ADHD, Asperger, dyslexia, Ritalin, Concerta, Adderall, Risperdal and Clonex are words that most readers will be familiar with, painfully familiar, if somebody in their family has been diagnosed with a specific syndrome or disorder and medically treated.

In this chapter, however, we will focus on those diagnoses and drugs not necessarily known to most of the population affected.

The clinical classification of our children into labelled categories could be seen as part of a conspiracy. The education system, the medical system and the pharmaceutical companies form an unusual and aggressive cooperative in treating our labelled and classified children.

The three organizations share a symbiotic interest. The education system seeks *teachers' silence* so it will not need to cope with children with special needs.

The pharmaceutical companies have obvious monetary interests and use the medical profession to assist them in marketing and promoting their drugs. Not surprisingly, most parents cooperate with them, as they think that if the paediatrician or the family doctor recommends certain treatment it must be for the benefit of their child.

It is important to emphasize that the author is not against medical treatment and in most cases medication is needed and justified.

The ambiguity and the broad spectrum of inaccurate diagnosis require investigation. This applies to most of the Western world, including, to a certain extent, the UK.

If we accept the current definitions of Attention Deficit Disorder (ADD) syndrome, then probably most of the population falls into this category. This alone will increase false treatments, which may have unnecessary and sometimes harmful effects.

Moreover, it seems that this wide classification into groups requiring medication is also in the interests of the newly created profession of caretakers. The proof is in the large number of psychotherapists, coaches and other so-called experts who have emerged in the last decade in a number of guises. To all that we can add the obvious interests of the pharmaceutical companies.

Obviously, there are cases genuinely requiring treatment and certain children will react positively, but there is concern about those who will be affected for life by wrong classifications and harmful medication.

The need to diagnose both children and adults and classify them in the appropriate categories is diversified. Legal, clinical and educational needs must be analysed and reported to the appropriate authorities. The increase in supply exceeds the increase in demand for the experts and institutions available and it seems that the entire population requires diagnosis.

It is essential to derive a diagnostic method which can minimize false positives, and especially to distinguish between sufferers of real syndromes and those who show certain signs of the said syndrome but are actually considered to be normal. Below are examples of cases which only experienced professionals can diagnose correctly.

- Evolutionary defects, seen in children with study difficulties such as attention deficit, may look similar to those evidenced in the ADD syndrome.

- Emotional situations that may cause attention problems, such as children under stress, abuse, neglect or other distress.
- Temperamental children maybe wrongly diagnosed as hyperactive although they are within the moderated hyperactivity range.
- Chronic disease, sleeping disorder, hearing or sight problems and malnourishment may cause misdiagnosis.
- Mental polarities such as genius or retardation may be interpreted as ADD.
- Behavioural disorder such as lack of motivation to study or insubordination may be interpreted as ADD.
- Fear or panic disorder that may be expressed as hyperactivity and attention difficulty like fear of failing in school.
- Depression disorder that is expressed in decline in activity, social disconnection and nervous and irritated moods is a case for misinterpretation.

The frenzied diagnosis and treatment applied by institutes and individual care providers is obvious from the many internet forums on the subject.

Worried parents visit numerous care providers, until they find the one who will supply them with the diagnosis that they seek. Given the large number of professionals available, they will certainly find one to support their diagnosis.

There is an endless and inappropriate war between school-appointed professionals and parents regarding the need to transfer children to the special education system. It exists everywhere where there are such options for the special education system but is particularly prevalent in the Western world.

The prejudice of such an education system, inflicting guilt and personal failure upon parents, means they try to resist the recommended change.

Ritalin, Concerta, Adderall, Risperdal and Clonex are only some of the drugs prescribed for our children by psychiatrists.

All this is in addition to vitamins, food supplements and over-the-counter drugs which are available without prescription and whose efficacy is questionable.

Ritalin, for example, is a stimulant which is very similar to cocaine. Long-term use can cause undesirable changes in a child's brain.

Dopamine is a neurotransmitter or a nerve conductive stimulant. It is a chemical substance in the brain that has a significant impact on attention and concentration. When concentrating, most children experience a substantial reduction in sugar and oxygen levels.

In those conditions of deficit, children look for excitement, which is sometimes dangerous, for the purpose of increasing their dopamine level.

The brain's need for extra dopamine makes children search for stimulating and exciting activities.

The best and the healthiest way to increase dopamine levels naturally is by controlled physical activity. Obviously, there are medications to achieve that target as well, but these are not recommended as a first resort.

Over-exposure to dopamine may cause hallucinations and visual phenomena, which are typical symptoms of schizophrenia. A very interesting phenomenon relates to the brain's need for dopamine. There is a clear correlation between children with special needs and their skill in music and drawing. We can explain that phenomenon by the extra dopamine generated during those pleasurable activities.

Ritalin and cocaine affect the dopamine system which helps to control functioning of the brain during experiences of pleasure. The subject of side-effects is very problematic, especially for long-term drug treatments.

Here is a somewhat graphic description of what is going on in a child's brain when s/he is receiving certain medication. The brain contains nerve cells or neurons.

Each one of those neurons generates connections called synapses. The connections are enabled by electro-chemical communication.

We know that the more we learn, the more those connections increase. In the case of brain hyperactivity there is an increased number of connections between the neurons that are responsible for certain actions or creative ability.

Certain drugs work as inhibitors to disable those connections. In other words, the drug prevents the creation of some of those connections.

It is unclear which connections exactly they act on and there are no clinical scientific proofs of effects on the creativity or other functions of the child if the drug were not taken.

In addition, it is not clear which new connections could have been created, or the short- and long-term effect. In certain cases, the drug is given for the purpose of lowering the hyperactivity of the child, so s/he will cause no disturbance in class or at home. We do not know if the use of such inhibitors means the loss of another Einstein.

Any new drug needs marketing and clinical research to prove its efficacy. Pharmaceutical companies employ physicians to assist them to this end. Because of their particular relationship with powerful pharmaceutical companies, physicians prescribe certain drugs that are not necessarily appropriate, which can lead to conflicts of interest in certain cases.

The cooperation and collaboration between the education system and the supporting psychiatric field are criticized by those who claim that drugs are too readily prescribed. This raises the question of whether our children's interests are being best served.

Is the need for classroom silence parameter factor in the decision to drug our children?

Obviously, there are many cases where drugs are required.

There is a need, however, to check and seek second opinions about the need for specific medical treatment, the need for and the efficacy of the specific drug and its correct dosage. In addition to psychological and medical treatments, there are many supplementary remedies. Not all have rational and scientific bases and the confusion is exploited by certain care providers for monetary gain.

Homeopathic treatments have been disputed and their efficacy is questionable, but if they relax children and their parents, then why not use them?

It is advisable, however, to consult the family physician first especially to avoid any undesirable drug interaction.

Certain parents simply will not accept the diagnosis and the fact that their children need medical, psychological treatment or special education. They run between the experts and care providers, until they find a diagnosis they can live with or they think is the right one.

Those parents are inflicting unnecessary pressure on their children and family. The impression often given is that they require treatment more than their children.

The rate of false positives or misdiagnosis is too high. It is too high even if we take into account that sensitivity together with specificity is above 80%. If your child falls within the 20%, for you it is 100% misdiagnosis, unnecessary and sometimes harmful treatments, which may affect your child's entire future life.

Children with syndromes have different communication systems from the rest of the population. Therefore, someone is needed who can translate their needs into a language that adults can understand.

In psychometric and other tests, children with ADD and other syndromes may face an unfair situation compared with normal children in two respects. One refers to the total time dedicated to the test and the other to the expected speed of responses to the questions.

In many cases children know the answer, but owing to the syndrome their concentration and response time may be longer than that allocated.

ADD and ADHD appear in boys about twice as much as in girls. Why?

Is it because more boys are diagnosed or because behavioural disturbances are less observed in girls? Does the reason lie in genetics or hormones?

In the middle of the twentieth century, the number of daily tasks and external stimulation were fewer than those of today and they moved quite slowly on the timescale. This is obvious in the films from that era. Even the action movies were relatively slow, but had substantial text and dialogue.

Today tasks have evolved to multi-tasks; films are faster and contain more visual stimulation but less text and dialogue, except in the films of Woody Allen or Ingmar Bergman. In today's society, whether it is democratic or not, the majority decides and expresses opinions, usually biased and loaded with prejudice, about who is exceptional or unacceptable.

It starts with kindergarten and lasts throughout our entire life. During the dark ages, we burned those deviants who looked odd, behaved irrationally or said strange and illogical things, like 'the earth is round'...

Galileo, Da Vinci and others were strange by normal standards. It is hard to describe what our world would look like without them. We made their life difficult and in spite of that, they achieved excellence and innovation in the arts and sciences.

For certain children, many daily sounds are translated after a while into noise. For some of them it is difficult to cope with all that noise. It is probable that the problem existed in the past but we failed to diagnose it, labelling such hyperactive children as naughty or slow.

The future entails more challenges and more noise, and there will be a need for children to face those challenges.

Many famous people like Edison, Mozart, Tesla, Dali and others were hyperactive. They were known for their unique temperament. They achieved greatness, perhaps because of rather than in spite of their attention deficit disorder

If the genes of ADD children have been selected or designed by nature for a new generation of creative inventors, then perhaps the phenomenon does not need treatment as much as refinement and acceptance. By treating such children we inhibit evolution and progress and we reduce them to acceptable, average and normative levels. Nature identified a need and it is trying to supply humanity with a new adaptable breed who will satisfy that need. They are the special *children of tomorrow*.

Accordingly, we should consider the qualifications and talents of those special children as a gift and not as a syndrome which needs psychological or medical treatment. We need to support and encourage them to develop those talents and not inhibit them. Such children have grown up to be creative and passionate adults showing a significant decrease of the negative phenomena defined by the experts.

Verbal violence has no less a negative effect than physical violence and both should be eliminated. Parental insults, yelling and threats achieve nothing apart from discharging their frustration. These situations cause the child to disconnect from his/her surroundings and withdraw into him/herself.

The child faces many difficulties at school and ought to feel that at his/her home s/he will be protected, understood and loved.

There is an obvious influence of family relationship on the condition and the functionality of the child.

It should be emphasized that nobody is guilty in this situation. It is useless to blame parents or grandparents for bad genetics.

Not only is it incorrect, but it adds to the bad atmosphere that the child has to live in.

There is a significant difference and distance between children and parents, much more than in the past. It is probably the first time in human history that children have special skills that their parents lack, such as IT and other technological skills.

This knowledge gap generates disrespect and a certain haughtiness in children, who are disappointed that their parents cannot operate a DVD or computer.

Our children are quick to comprehend, impatient and intolerant and will reject deviants. It is true that children were cruel in the past to their eccentric friends, but these days their cruelty is amplified. Today's children are quite superficial, searching for the easy, simple and immediate or instant.

Children born into the middle-upper class dedicate less thought to future economic and financial issues, and the consequence is a low common denominator and lack of effort to achieve for excellence.

The professional literature recommends a *strong father*, but not a violent one, who can generate proper parameters in terms of permitting and forbidding certain behaviours.

Here is a short list from those recommendations, about ideal parents for children with ADHD syndrome:

The boy has need of a strong and strict father.

*The father may be seen as 'ghost terrorífico' by the boy, who
may develop fear and hate,
but at the same time also love and respect.*

*The presence of a strong father is indispensable
so that the boy will be able to process the hatred
that inevitably he will feel toward someone.*

The actual presence of the father at home is indispensable to support the maternal authority and to be constituted as the representative of law and order.

The attitude of the mother is fundamental, together with that of the father.

A problem arises in cases when a significant difference between parental profiles is obvious.

The child suffers distress when s/he lives with a dominant father and with a weak mother or vice versa. Psychologists require such parents to coordinate their positions and attitude.

The dominant parent will be required to adopt a lower profile and the weak parent a higher profile and be more assertive. When one of the parents is unable to perform that task, the situation may get worse.

There are cases where there is a gap between the parents in the permissive area. This gap is usually generated by the mother, and the authoritative figure is usually the father.

In such a case it is recommended that the father should stop being *responsible* for the *punishment* issues, which will lower his profile. It is essential, however, that at the same time the mother *increases* her profile and becomes more *assertive*.

In the past, problematic children were 'treated' with 'local punishment' by their parents and teachers. In Europe, they used to make children kneel on corn, hit their fingernails, slap their face and inflict other physical punishments on different body parts. Parents used to accept these punishments, albeit with some reservations, and to support their understanding they sought reassurance from the Bible:

He that spareth his rod hateth his son; but he that loveth him chasteneth him betimes (Proverbs 13:24).

This means that a father who does not discipline and punish his child is not educating him and therefore he does not love him. Although the reference to 'son' is to the generic name for children, in this context boys are the subject.

Punishing children using physical violence never solved anything; on the contrary, it only amplified the conflict and generated additional problems.

But…yes, there is a but…, society went from one radical extreme to another. Although today such punishment is legally forbidden, it still exists. We went from being punishing parents to being soft and over-protective. Today's parents are spineless, and fail to set any limits, so that everything and anything goes. This does not reduce the level of violence but actually increases it.

What happened? Why are we a more violent society than in the past?

Obviously, there is no one simple answer to that question. One reason might be that we opened the door unwittingly and let the violence in.

We did it by misunderstanding our children, without establishing proper family relations and by ignoring their needs. We let them call us by our first name and not Dad or Mum, we became their friends, instead of keeping a proper distance. We lost the discipline of the relationship we had with our own parents. In the context of parents, the high divorce rate among the general population, and particularly among families with special children, should be mentioned as a factor affecting children with certain syndromes.

Parents in the past spent more quality time with their children. Research shows that today parents spend no more than fifteen minutes on average per day.

Because of their agonized consciousness of the difficulties, some parents became servants of their children and lost their dignity.

The education system in school faces many problems.

There are no leadership, no rules, no discipline and no order. They have no solution to the increasing rate of violence.

It should be emphasized that violence is not genetic, but it is acquired, and there are many ways to minimize it.

Real-time online video surveillance with the awareness of the children can reduce violence at school. Such monitoring, even at the cost of privacy, is essential to document and monitor visitors and strangers in the school and surrounding area. Drivers drive more carefully when the police are nearby. We assume that children will behave better when they know that they are being watched.

There is a significant increase of alcohol usage between the ages of eleven and seventeen. There is no doubt that it is one of the reasons for the increase in violence. The solution is similar to what Mayor Rudy Giuliani implemented in Manhattan, New York. He used 'zero tolerance', imposing immediate punishment, even for small crimes, and succeeded in significantly reducing New York's crime rate.

Alcohol and medication are a lethal combination. Special attention should be paid to those children who use alcohol and drive. Investment in advertising at the national level, especially in schools, using audio-visual and real lectures, may assist in reducing both alcohol and violence levels. The campaign must include strong messages about the results of using alcohol and the punishment imposed if violence is used.

Parents must seek support and cooperation from their own parents. The grandparents of the child are part of the rehabilitation and treatment process.

There is a need to verify that the child is not mistreated by the grandparents because of their ignorance of the situation. Their cooperation is essential in carrying out the policy and the treatment recommended by the proper medical and psychological care providers. Parents must explain the new situation to the other children in the family, give them specific tasks and obtain their cooperation in treating the sibling with the syndrome.

It is vital to explain that their sibling is not sick, but needs help. S/he has unique qualifications in certain areas and lacks them in other areas. S/he needs their love and support, which s/he will reciprocate in future.

We see an increasing gap between the relationships of children and the education system. The schools of today are quite archaic. In most cases they have not followed the changing technology and the changing needs of modern children. This statement is supported by the archaic education methods and the outdated tools used.

In the past, schools not only taught but educated as well. They were almost the only source of knowledge and information, and therefore it was impossible to criticize them or verify that the information taught was correct and up to date. Today children have to sit quietly in one place, listening to the teacher's voice, which is sometimes quite loud and eventually becomes 'noise' in the brain of the special child.

On average, there are about three disturbed children in every classroom, some of whom have ADHD syndrome. Instead of removing or suspending them, another method is advised. Collaborate with them, let them participate in class activity, give them the stage and flow with their needs. Usually, children who disturb in class send a message: 'I am bored'. This message is directed at the teacher. Teachers should identify the problem, i.e. why those children are not involved in class activities.

Often the solution is simple. They were teased or they simply did not understand the question or the situation; it is important to find out what bothers them. Removing those children from class may help temporarily but the problem is not resolved but often intensified.

By removing children from class, teachers signal to the other students that they have marked those children as problematic and they will treat them accordingly.

Yet if teachers succeed in persuading them to cooperate, they will benefit from that as well. They might become better and more acceptable teachers, admired and appreciated by the students, the school and especially by the parents.

Teachers are recommended to develop good relationships and cooperation with parents. This is important in general, but particularly in cases involving children with ADHD or other syndromes.

In the regular education system, teachers do not have the patience and some of them lack the qualifications required to deal with such children. Empathy and cooperation between teachers and parents in certain cases may eliminate the need to transfer the child to the special education system.

Imaginary friends are part of a normative evolutionary phase which gives the child an efficient platform to express him/herself, usually in a very creative way. Most children have imaginary friends at some point in their early life. Those friends will eventually disappear. They disappear simply because with time the child loses interest. For some children, the imaginary friend may appear in the form of an animal, such as an elephant, tiger or dog.

The child believes that the imaginary friend will fulfil his/her wishes and believes that the friend is a kind of 'super-friend'. This friend is very efficient and effective in the child's life and may have many goals. It may help in facing and coping with feelings that are difficult to express.

Sometimes, those feelings are negative in nature. When the child is angry and wants to react with violence, s/he is aware that it is forbidden, and then the imaginary friend comes to help.

All negative feelings and wrongdoings are transferred to the imaginary friend: 'I did not do it' or 'He did it'.

Because the imaginary friend was created by the child, s/he feels safer and relaxes with it and the communication between them solves many problems, including loneliness and boredom.

The friend eases the child's pain, anxiety and fear and, as mentioned earlier, eventually he will disappear. It is important to remember that the imaginary friend will not replace real friends.

*What cannot be resolved with love
can be resolved with more love*

The following methodology may be applicable not only for ADHD cases, but also for cerebral palsy, deafness etc., where communication or the lack of it is a dominant factor in the treatment process.

The proposed treatment and methodology principally involve the inner family circle, where we all have ability to implement the recommended changes.

The regimen sits apart from the conventional methods, which constitute a kind of general formula applied to diverse classified groups as a collective treatment.

Assuming that you the reader have somebody in your family with a syndrome and you wish to assist and get involved in learning how to treat and improve life for that child and his/her family.

The subjects described here will enrich your tool-box and will enable you to tailor the treatment to your case. The recommendations made here as part of the proposed treatment must be seen as a complementary activity to the diverse treatments proposed and implemented by qualified experts in the field.

It is important to perform the diagnosis as early as possible, when irregular behaviour is observed either by the parents or by the education system.

In order to understand the special situation that your child is in, imagine that you are transferred to a parallel universe, where you do not understand the local language.

You are in your own house, surrounded by a worried family, everybody is crying. You are lying in bed; you do not understand what's going on. You are shouting, but your voice is not heard, you want to move, but you are paralysed. You are in a nightmare and totally helpless.

Less dramatically…you are in a class surrounded by children. You are the same age as your child, but the other children speak a language which you do not understand. Your entire communicative ability is limited to touching…

In a short time you will be outcast, labelled and alone. In your mind you are fully aware of the surrounding situation, but you are unable to communicate properly.

This scenario is similar to the situation your child is in. They need to learn and adopt the proper communication skills. Some children will use physical contact, some will try to attract attention by shouting and cursing and some will simply disconnect, give up and leave the class or the house.

The following actions may assist you to understand better and become closer to your special child.

First, get acquainted with your child, without any prejudice. Try to ignore what was said, diagnosed and labelled by all professionals and others. Invest quality time and monitor your child's activities closely. Find the things they like and do not like in all areas: food, play, school, study, friends, etc. This information is critical if you are to arrive at the right treatment.

When a child loves and enjoys a specific activity, positive and efficient communication exists during that time. Use that time to merge or to insert messages and desired behaviours that you would like the child to adopt, improve, or correct, without interfering or disturbing the specific activity.

The child will not communicate well when s/he is required to do certain tasks against his/her will. If forced to do it, usually s/he will cooperate, but without any enthusiasm. This is not a good time to educate or send corrective messages.

Let us assume that the child is particularly fond of the image of Tom the cat (from the *Tom and Jerry* TV programme). S/he will be more open to receiving messages from a puppet in the shape of Tom than from a parent. There are many suitable interactive programmes and a list should be available from your therapist or on the internet.

Food has a significant effect on the level of attention disorder. Therefore, it is vital to monitor your child's diet. It is sensible to avoid junk food, preservatives, edible colours or over-indulgence in snacks and sweets. You should be aware that most cereals comprise over 30% sugar. Although sugar's influence on ADHD is a controversial subject, it is currently recommended to avoid excessive usage.

It is important to provide a proper breakfast containing proteins and carbohydrates such as whole-wheat bread and egg. Right nourishment at the right time of the day is important, especially for those who are on medication. It is important to eat before taking the medication, especially Ritalin, which causes loss of appetite inter alia.

Proper and timely sleep has an influence on the child's functioning and behaviour. Therefore, a routine sleep of at least eight hours is desirable, as is sleeping at fixed regular hours and not after 22:00.

Special children have a hearing capability and sensitivity to noise above the average and most of them like music. Therefore, it is important to facilitate a quiet and relaxed atmosphere.

Do not do – Do not yell, insult, curse. Do not be judgmental and never raise your hand to a child.

Do – Love, hug, encourage, support, forgive, understand, smile, and love more.

You should award and encourage a child for good behaviour and revoke certain rights for inappropriate behaviour. You have to explain to the child the basic law of physics:

For every action there is a reaction.

Therefore, 'If you won't eat, you'll be hungry', 'If you hurt your friend/sister/brother, they will feel pain, suffer, and they will not be your friend and will not play with you', 'If you don't shower, you'll smell and your friends will avoid you'.

It is important to respond immediately to any violent expression and monitor closely what is going on at school. Whether your child is the abusee or the abuser, you need to respond and treat the problem from the first with 'zero tolerance' Participate in parent-teacher meetings to resolve disputes between children. Without immediate and proper intervention, the situation might get worse.

If the child is on medication, it is vital to look for possible side-effects and report them to the family physician. Many drugs are addictive and the right dosage is no less important than the right medication. If you observe anger outbreaks, and the child is constantly tired and difficult to talk to, it is important to check the medication and its dosage.

When you see that your child is capable and enjoys doing a, b or c, but does not like or is unwilling to do d, e or f, you should focus on a, b or c, and compliment and encourage the child. Later you can teach him/her slowly, step by step, d, e or f, but try to get some cooperation and do not force the pace.

You should avoid 'angry prophesies' such as: 'If you do not...then you'll be this and that'. In general, do not send negative messages or generate negative energy. Your child is in a difficult phase in his/her life. S/he is fighting to be accepted as 'cool' at school and at home s/he expects warmth, love and support.

When s/he has to listen to 'prophesy speeches' too often s/he may give up and submit to despair. Sometimes, a prophesy is self-fulfilling.

When a label or certain adjective is used, the child may adopt it, even if s/he does not necessarily deserve it. S/he may say to him/herself: 'If I am expected to be...then why should I try being something else?'.

It is advisable to teach the child to take care of him/herself. Yes, it sounds unrealistic, but it is possible. First, show him/her, in a video for example, another child who looks like him/her and has a similar background. That child will carry out certain tasks, and will talk about and explain them in a way that your child can understand and relate to.

An alternative might be to video-record the child unawares and show him/her later. You can explain that this is how s/he behaved, said or did and together you can find what was wrong and what would be the right way the next time.

The process of training the child in diverse simulations of situations will persuade him/her that there are other ways of self-expression. Accordingly, step by step s/he might change his/her pattern of behaviour and adopt a more acceptable one.

Physical activities such as judo, horse riding and the like are strongly recommended. Such organized group or individual activities may enable the child to divert their extra energies and release them at a suitable place and time.

The child may adopt rules and disciplines through appropriate activities, self-organization, persistence and the carrying out of certain tasks.

In addition, it is desirable to organize literary activities, such as discussions about books, poems or songs, which may assist the acquisition of self-confidence. In general, special children do not like change. Therefore, you should prepare a weekly schedule, which is more or less filled with known, fixed and mutually agreed activities.

Children are quite conservative and will not try new things willingly. For example, usually they will refuse to eat new and different food. In this context and in general, it is recommended to adopt the following attitude or rule: 'You don't have to eat it, but you have to taste and try it'.

This approach is applicable to many contexts; it will encourage curiosity and experience of new things, but it should not be enforced.

In terms of negotiations, what is good for adults is also good for children. We as parents should emphasize the areas and the subjects where we have agreement and mutual interests. We should use our parental authority to emphasize the rationale and logic behind our demands, without forgetting the benefits, namely, the reward for cooperating or the consequences of disagreement and unwillingness to cooperate.

The following practical advice is aimed at parents of special children.

Keep your promises and do not make any if you have no intention of keeping them. Tell your children about your childhood. Spend quality time with them as often as you can. Be an example and a role model to your children. Take holidays, go on trips and spend time on a one-on-one basis. Tell them stories; they will remember your voice more than the content of the stories.

Take them to your workplace. Get to know their friends. Always listen to them. Make a kite and fly it together. Define consistent rules and limits. Take them to museums, zoos and other fun places. Make them participate in home tasks and various assignments. Most importantly, show empathy, warmth and lots of love.

Parents have different styles: permissive, dominant, authoritative, indifferent or even, sadly, neglectful and abusive.

The desired style is an efficient attitude which enables one to find a balance between respect and consideration for the child's need and parental expectations.

There should be a balance between excessive parental presence and the lack of it. We should provide enough freedom and privacy and at the same time monitor and supervise the child's activities.

During adolescence the child goes through many changes. Those changes are emotional, physical and social, all in a relatively short time.

This is a difficult period not only for the child, but for the parents as well. Parental authority is questioned and power struggles are quite common. Parents should learn to handle conflicts and confrontation wisely, decisively and with lots of love and understanding. You should adopt a reciprocal basis of give & take and reward & punishment and let the child win some arguments as well.

Parents should set clear consistent standards and be a role model to follow. Standards should be such that parents themselves can stand by them and follow them. You cannot expect total agreement on all objects of dispute. As in any other conflict, you may insert certain 'fake' demands for the purpose of removing them at a later stage. A conflict will be positively terminated when both sides feel they have won.

When we set demands and standards to follow, there is a need to respond properly when they are not followed or are violated. Specific standards are culture-oriented and might be problematic, especially in a multi-cultural but highly segregated society. The activation and implementation of parental authority is essential for the proper evolution of the adolescent child and is particularly important for the special child.

If the child is experiencing a difficult day, if s/he was bullied at school or just in a bad mood, which is a common situation, try the following experiment.

Draw a circle of about one inch diameter and blacken it on an A4 sheet of white paper. Put the page in front of the child's eyes and ask: 'What do you see?'.

The usual response is: 'A black circle'. Move the page a few feet away and ask: 'What has happened to the big black circle?'. 'It became a small point': at greater distance the point is barely visible. Explain to the child that when the big black circle is close to his/her eyes, it is equivalent to his/her current feelings, namely, everything looks black and gloomy. But in time it will become a small point, which will disappear and all will be white and clean.

Special children are more visually than textually oriented; therefore such a metaphor might be accepted and understood better than verbal explanations and encouragement, which should be given in any case.

Children at school have always picked on the weak, the odd and the different. It is a cruel reality, even in the animal world. They make fun of others, ostracize them and sometimes even use physical violence.

Your child is in a difficult situation. If s/he responds with violence, s/he might be expelled from school. If s/he does nothing, s/he might become the butt of the class. There are no magical solutions, but you can minimize the bullying scenarios by using the following process.

Summon as soon as possible after the event a meeting under school's auspices with the parents of the bullying child and parents of other children involved. It is important to prepare the parents before the meeting in order to minimize the possibility of any dispute. It should be made clear to the bullying child's parents that lack of cooperation on their part might cause their child to be expelled from school.

The meeting could be held on school premises, but it is better to select a neutral place like a garden or a park or somewhere without external disturbances.

Sometimes, one meeting is enough to show the children involved that their parents are cooperating and share a mutual interest in resolving the situation.

Moreover, the children must see a united front on the part of parents and school. They should realize that they unanimously present acceptable, expected and desired normative behaviour.

The special child should be taught not to raise his/her hand, but at the same time not to run away. S/he can confront the bullying child face to face, tell him/her that s/he will not tolerate abuse, s/he does not expect such behaviour from a classmate and if it continues s/he will be forced to report it to the teacher and parents.

Such confrontation sends the bullying child a message, 'I am not afraid of you', which may cause him/her to think twice.

You are sitting in your car. What are all the options open to you? Move forward, backward, left, right. Yet there is one more option, which is to stay where you are.

Teach your children that when they are faced with choosing between a or b, they always have a third option, namely not to choose or not to decide now. 'I'll think about it', 'I'll consult with my parents and choose later' should be their response. In life there is more than one option, regardless of what others may wish us to do.

Children make many decisions every day. Some of them are normative and others are a response to threats made by hostile classmates or other children: for example, 'You have two options, either you give me your sandwich or I'll hurt you'. You should teach your child that s/he has more options when s/he is given two bad choices. I mentioned earlier the law of physics governing action and reaction. In this case, the robbed child could respond: 'If you eat my sandwich, I'll be hungry and I'll have to inform my parents, who will call your parents', 'If you hurt me, I'll suffer pain and I'll be forced to report it', 'I have an idea; today, we will share my sandwich and tomorrow we will share yours'.

A stronger child might respond: 'I have no intention of giving you my sandwich, and about hurting me, well, you know what the result will be'.

If you are the bullying child's parents, do not use any punishment, because it might only make the situation worse. Invite the other child for a joint picnic.

Try to connect and not separate children. Explain to your child the consequences which all of you will have to suffer if s/he does not behave.

There are parents who still say: 'If he is hurting you, be a man, hurt him back'.

This attitude not only will not resolve the conflict, but in most cases it will enhance it. Violence triggers more violence.

The phenomenon of bullying and violence is much less common in the special education system. One of the reasons may be because there are fewer violent interactions between children at special schools. The qualified teacher will handle any outburst of violence immediately, firmly and sensitively.

Violence is an outburst of energy without control and aim. It is a waste of energy, inefficient and useless. Therefore, we have to channel our children's energy positively. There is no point in suppression and punishment in such cases and education on the right use of such energy must start at an early age.

A punishing parent is a powerless and weak parent. This parent has no practical solution and the only response s/he can make is to demonstrate his/her parental authority by applying punishment.

Many parents respond with a negative attitude: 'Don't go there' or 'Don't do that'. They tend to emphasize 'what not to do' and not 'what to do', which is the preferred positive attitude.

We have to teach our children to take responsibility for their actions and not let them blackmail us emotionally. They can do it with some success as they are aware of its effect and realize that it works.

A positive and a rewarding daily interaction system between us and our children should be developed. We have to show involvement as well as supervision and reward good behaviour with compliments, without generating a link between a child's achievement or success and monetary or material compensation.

It is preferable to say to a child 'You succeeded because you have invested a lot of time and effort in preparing your homework' rather than 'You succeeded because you are smart'. The latter sentence sends the message 'I am smart, therefore I don't need to make an effort to study', whereas the conclusion to be inferred from the former sentence is 'I succeeded because I have invested' and this will teach the child that in order to succeed s/he first has to invest effort.

In the case of failure, there is a need to encourage children in a positive way and explain that: 'It is OK and normal to fail occasionally, it is part of life' and the question is how to face and cope with failure and rejection. Teaching children to strive for excellence is good but at the same time it is important to prepare them for occasional failures as well, to encourage them to 'climb back on the horse' in case of a fall or failure.

Success after failure is satisfying and will strengthen the child, who may learn from it for the future.

It is important to emphasize that the aim of this chapter was to generate stimulation to find creative ways to assist ourselves and our children to enjoy better communication and a better life together as a family.

8. The Philosophy of Negotiation

It is fair to say that we have been negotiating since birth. We started as children, negotiating with our parents, family and friends.

When we were given options such as 'if then else', we tried to improve the conditions by negotiating a better deal. Occasionally, we even negotiated with ourselves.

Our ancestors negotiated with God at the creation and later with everybody else. We have improved and expanded our negotiating skills and today they are widely applied, not only known and practised by attorneys. Negotiations are conducted in diverse arenas, each requiring different negotiation skills, but we can find commonalities among them.

We need to negotiate with terrorists, kidnappers and bank robbers, where the outcome can be deadly, and we may save lives if we succeed.

We negotiate in shops and markets, where no sale is the worst case scenario.

There are mild and aggressive negotiations. We negotiate contracts for buying and selling and 'if then else' is one of the basic elements in negotiation.

One-sided ultimatum negotiation is common as well, such as: 'These are my terms', 'Take it or leave it' and 'It's not negotiable'.

The philosophy of negotiation takes into account that the parties are interested in executing a common deal.

Each party, however, sets certain standards, conditions and limits which they will accept or agree to.

For example, *A* has certain goods for sale and *B* is interested in purchasing. Here are several scenarios:

1. The goods have a fixed price set by *A*, like drugs in a pharmacy or goods in certain stores. *B* has no negotiating options; either he agrees to the price and buys the goods, or refuses to pay the asking price and goes shopping elsewhere.
2. The goods have a fixed price set by *A*, but there might be a special discount. The discount may be subject to submitting previously published coupons or based on certain conditions, such as store membership or seasonal sale. In this case, the buyer is given a conditional or unconditional price discount, but without negotiating options.
3. The goods have a fixed price set by *A*. The price caters for a certain discount, based on the buyer's negotiating skills. Buyer *B* has the option to offer a lower price and *A* in return may either accept or make a counter offer, until they reach agreement; otherwise there is no deal.

We will analyse a general negotiation case between seller *A* and buyer *B*.

The obvious goal of *A* is to sell for the highest possible price, and the opposite goal of the buyer is to pay the minimum price. The assumption is that both are willing to enter into the negotiation phase to conclude the deal.

We will try to understand the philosophy behind their thoughts and strategies for optimally achieving their goals.

The seller knows that he is entering into a negotiation arena and accordingly has embedded in the asking price a certain acceptable discount margin.

In his mind, the seller may have set a minimum price that he would accept, and below that minimum there will be no sale.

A has set his asking price at $100, but he realizes that the chance to sell at the asking price is quite slim. He has set a minimum selling price a priori at $60.

Buyer B may have similar thoughts. She knows that the asking price has a certain amount of discount, which is subject to negotiation. Therefore, she decides not to agree to the asking price of $100. In her mind, she decides to pay a maximum of $80.

Seller A encourages buyer B to make an offer. B tries a 'fishing trip' and provocatively offers only $50.

Hearing the ridiculous offer, A overcomes his inclination; he wants to sell, but will not accept $50 and he responds with a counter offer of $80.

B is hooked now, as she realizes that the new offer is the price she is willing to pay. Although she may have the option to end the negotiation by accepting the new offer, B as an experienced negotiator makes another attempt and offers to pay $60.

Seller A realizes that they have reached the sum he is willing to accept; he has the option to terminate the negotiation as well, but A is also an experienced negotiator so he uses the 'meet them halfway' system. This is a classical negotiation method, where both parties want the deal and each thinks that the middle of both offers is a fair way to close the deal. The parties agree the deal and the negotiation closes for $70.

In this case both parties are happy, the buyer who thought she would have to pay $80 got the goods for only $70, less than she expected, and at the same time the seller, who thought he would be willing to sell for $60 actually received more than he anticipated.

A good and a fair deal is when both parties end up happy with their decisions. Both parties made their optimal choices and both will leave the negotiation with a winning feeling.

Obviously, this is the ideal scenario and it has many different endings in terms of the closing price or in walking away from the deal.

In an auction sale, the final price is set by the auctioneer's hammer and no direct negotiation is possible. The process of bidding from the starting price until the hammer comes down involves a lot of psychology, however.

Potential buyers are bidding, or more precisely fighting, among themselves, in order the get the item they desire.

Often they are carried away by their emotions and pay more than they decided a priori to pay, or even more than they can afford. The auction hall is the ground for the hunt.

Ego, usually male ego, is one of the parameters which will decide a final price many times above market value or the buyer's real need.

'There is a cheaper item over there', said the potential buyer to the seller. The seller asked for $100 and the buyer said: 'But over there, they asked only $70'. The seller replied: 'OK, so buy it there'. The buyer said: 'But they have sold it already'. The seller said: 'Thanks for the info, so now my price is $120'.

Another version might be: 'So why don't you buy it over there?'. The buyer: 'Unfortunately they have sold their entire inventory'. Seller: 'OK, when my inventory runs out, I'll sell it for only $50'.

Pricing is a science with a lot of psychology. When the price is too high, people will not buy; too low a price is not only loss of extra profit for the seller but it is not appreciated by the buyer.

Cheap is cheap

Pricing is affected among others by context and location. The same item sold in a market, in a small shop or in a boutique in an affluent neighbourhood can make a significant difference in price. Price is obviously affected by the uniqueness and the rarity of the item.

A rare nineteenth-century silver and enamel object made by Fabergé may fetch a significantly higher price than a similar item made elsewhere in another period by an unknown maker.

In an open trading market environment, price is often not displayed or fixed by the seller. Experienced sellers do not label products with a price if they are willing to negotiate. They will however make an ad hoc decision, based on buyer's image, origin, sex and other factors and set the initial asking price accordingly.

It is interesting to observe the diverse negotiation skills and methods deployed worldwide.

There is culture-oriented negotiation such as in the Middle East, where there is a significant gap between the asking and the final selling price.

Price may be affected by the gender of both the seller and the buyer, whether he or she is local or a tourist and even by appearance, smell, voice and other factors.

Don't ever bid against yourself

If you as the seller have set an asking price, do not start to change it when you see and feel that the buyer is not interested. If the buyer is truly interested, s/he will enter into the negotiating arena. In this case, if you lower your initial asking price, you will lose the starting negotiation point. Ask the buyer to make the first counter offer to your first asking price. This case is clearly demonstrated in the example described above.

The more you seem to be eager to sell the less you'll succeed

Do not run after buyers, play it cool. In most cases, you cannot persuade a potential buyer to buy if s/he is not interested. The buying and selling possibilities available on online trading websites such as eBay are: *fix, auction, make an offer* or *buy it now*.

The 'fix' price is the situation of 'take it or leave it' with regard to the indicated fixed price.

The 'auction' option requires one buyer to bid for the starting auction price and it may increase if other bidders are interested to buy; eventually the highest bidder wins.

In the 'make an offer' option, the seller enables the buyer to make an offer. In this case, the seller may accept the offer or make a counter offer, until they reach an agreement.

The 'buy it now' option may be combined with the 'auction' option, whereby the buyer may bid and compete with other potential bidders, or accept the 'buy it now' price and win immediately.

Negotiation skills are required in conflict management, arbitration, conflict resolution and mediation. If A has a conflict with B and A is our client, the best advice to A is to bring a third party C into the equation.

This may yield a stronger and a better result. By bringing C into the conflict, we consider future situations and relations that might evolve after the negotiation is terminated. It holds true especially in cases where A has to continue personal or business relations with B.

Negotiation is quite often seen as confrontation. Effective negotiations need not be confrontational, however. Setting the mood as aggressive and seeking to win means that there must be a loser.

The correct attitude of the opposing parties should not be to win the confrontation but to find a mutually agreeable solution.

It is necessary to control our emotions during the negotiation process. The more we lose control and become emotional, the less we will be able to achieve an efficient, desirable and mutually agreed solution.

We must make an effort to focus on the issues in hand and not on the specific and sometimes annoying personality of our counterpart. Blaming the other side is a definite distraction and an unproductive one.

One of the most important factors in efficient negotiation is to research and understand the needs of the other party.

To find a mutually agreeable solution to the problem, we need to assess the gap between our needs and any disagreements. We will be able to do so only after understanding the needs and worries of our opponent.

A typical example of how such understanding can be effective is the following scenario. Let us assume that two people have found a coconut and each is claiming it should be theirs.

You happen to be there and you are chosen to be the arbitrator. What would be your ultimate solution to this conflict?

Most of arbitrators would simply suggest splitting the coconut in half. In this case each claimant will have only 50% of what they want. Is it the best ultimate solution? Well, not really.

If you had talked first to each party to learn their needs before making your ruling, you might have found out that one of them is an artist interested only in the coconut shell for carving, whereas the other just wanted the milk and the coconut meat.

With this information you would be able to satisfy 100% of each party's needs and reach a classical win-win situation.

Timing is everything. Negotiations, like many other things in life, are time-dependent. There are better and worse times and places to conduct negotiations. When entering into the negotiation process, we should be prepared, learn about our opponent, prepare alternative solutions, not waste time on futile arguments, never get emotional and present persuasive arguments to support our claims. Essentially, the aim of negotiation is to cause a change in our opponent's perspective that may lead him/her to agree and come closer to our needs and desires.

A good and efficient negotiator is one who has the ability to persuade. Sometimes a negotiator needs to use his/her persuasive skills for the sole purpose of encouraging the parties to open up, to talk, to listen and to start a dialogue.

There are many persuasion techniques that are used by negotiators. Some are positive and some are negative. Among the positive techniques are basic physical touch and gestures like handshake, smile, compliments, respect and other small-talk to generate a certain atmosphere of familiarity.

The negative attitude used in the persuasion process may contain obvious or hidden threats such as 'if then else'. Monetary threats are usually very effective tactical methods, especially in financial arguments. Failing is another negative technique, and can be considered as a major psychological punishment. Therefore, when entering into the negotiation process, we should set aside our fear of failing or losing.

One of the basic and most primitive human motivators is fear. The fear of losing property, wealth, a game or anything else puts us in a weak position in the negotiation process.

We should try to control and monitor our non-verbal signals. Our body language sends out revealing messages, especially when we are angry, frustrated or eager to accept and agree to a proposal or to a compromise.

In our daily life we face conflicts that require negotiation in order to resolve them. A typical example is conflict between employees and employers, Employees want more money and better working conditions and employers want to make more profit and minimize their labour costs.

In certain countries and in specific cases, employees are united under a local labour union responsible for the actual negotiation with the employer. Such negotiation will normally take several rounds of meetings, in which different scenarios emerge.

There are cases where the union may demand certain changes, such as salary increases. There are several negotiation styles and methods that can be used either by the union or the employees acting independently.

They may enter the negotiation arena with an ultimatum such as: 'If then else…' or alternate demands and response may take place between employees and employer.

Experienced negotiators will add to the basic minimum demands an extra demand or several demands that will be given up later, as an act of goodwill, so that not all demands will be met.

Consequently, at the end of the process, both sides may declare themselves the winner. This mutually agreed compromise is essential for the ongoing relationship between the parties as they have to continue to work together.

In an aggressive ultimatum environment, employees must understand perfectly the philosophy and the needs of management, as they might not react well to an ultimatum. This holds true especially in cases where the employer is seeking a way to close the business, replace certain employees or perform reorganization.

Employers may use this ultimatum situation to resolve their business problems and get rid of trouble-makers.

Here are several scenarios between union and employer:

Union: 'We want a 20% salary increase'.
Employer: 'No way!'.
Union: 'OK, so what about 10%?'.

Such an early retreat by the union obviously signals to the employer that they are weak and they may end up with no or a small salary increase of maybe 5%.

Union: 'We demand 20% salary increase, otherwise we quit'.
Employer: 'OK, so quit'.

In this case the union were not prepared for such a response and they are now faced with the problem of how to proceed.

If the employer is just testing how far they are willing to go, there might be some room for compromise.

If, however, the employer is serious about closing the business or replacing certain employees, then the union may have caused irreversible damage to the employees.

Union: 'We demand a 20% salary increase. Our demand is based on the increase in the company's profit made during the last two quarters. This profit was enabled as the result of the extraordinary dedication, extra hours and production improvements of the employees. Moreover, in order to make it more acceptable, we are willing to accept this increase in two instalments'.

This is a reasoned claim by the employees and it is probable that the management will accept the principle and agree to some or even all of their demands.

Employer: 'As a result of your inability to supply the ordered goods on time, we lost our client. This caused substantial financial damage and I have to make a certain reduction in labour. You the union must supply the names of ten employees to be fired'.

Union: 'We understand the need but not the means to resolve the problem. We suggest that we do not fire those ten employees, but agree to make up the equivalent cost in salary reduction. This reduction will be in effect until the company is back in track. When the company is profitable again, we ask that the employees are remunerated with that salary reduction plus a bonus'.

In this case a reasonable employer would accept such an offer, as the benefit in the long run is obvious. The chief asset of any company is its employees. In this case they want to stay, accept a salary reduction and make an effort to succeed and put the company back on track.

The union was very clever to make such an offer. No employees will be fired. The loss in salary reduction may become a good investment for the future. Not only will employees get that amount back, but they will receive an additional bonus as well. It is almost a win-win situation.

Time-dependent negotiation is when employees enter negotiation after or during a strike, while the business has to deliver certain goods already paid for, for example. Each day of the strike costs the employer money and they are unable to deliver the goods they might well lose the order.

In such a case, the employer may have a monetary interest in ending the strike as early as possible and the employees know that. They also know that if they fail to deliver the goods in time they might lose their jobs. In such a case, both parties have an interest in ending the dispute as soon as possible and both realize that they have to compromise.

Both employer and employees can evaluate and calculate their win or lose time-dependent costs. The balance in most cases is somewhere in between the demands and the offers of the parties involved. Delaying the compromise or stopping the strike can be beneficial or advantageous to either party or to neither.

Obviously, there are many more variations of employer-employee relationships and they may contain many different elements depending on the type of business conducted. Employers' dependence on employees in a high-tech company is different from that on employees in a purely service-oriented company; the first have vital information about the company's products, whereas the latter can be replaced without loss of knowledge.

Replacing a high-level engineer is more difficult than replacing a hotel bellboy. Such a dependency plays a major role in the negotiation between the parties concerned. A business structure where certain key employees hold the management hostage by threatening to leave or go to the competition may put the management in a weak negotiating position.

The case may be even worse when those key employees hold exclusively the know-how technology.

The above-mentioned negotiation scenarios are applicable in certain cases to disputes between countries as well.

During the writing of these lines, Iran and the Western world are in dispute about Iran's plans to develop nuclear weapon capabilities. This is not the place to enter into the complex political and international relationships involved, but we can see that this conflict poses a significant danger to the region and to the world.

The Iranians want to upgrade their international status and image by joining the nuclear club. They might have achieved their goals without substantial objections, if they had had a serious and responsible Western-type democratic regime. Their leader Mahmoud Ahmadinejad, who constantly threatens the West, particularly Israel, is largely responsible for the grave international objections to their plan to develop and own nuclear weapon capabilities.

There have been many meetings and negotiations between the Iranians and several UN bodies, including the International Atomic Energy Agency (IAEA), in an attempt to resolve the dispute.

The Iranians started by seemingly cooperating with IAEA, but continued their plan to achieve their goals. Later, the pressure was increased so they stopped cooperating and started to make threats.

Ahmadinejad's statement that 'Israel must be wiped off the map' led the Israelis to react aggressively, including threatening to attack Iranian's nuclear plants.

Given the current Iranian attitude, this dispute cannot be resolved in an amicable manner. Israel and the USA are demonstrating by means of counter-threats and military exercises that they are willing to attack Iran's nuclear plants.
The Israelis, in the light of their past experience, see Iran's plans as a threat to their existence.

This is a classical case in which a lot of political issues and personal egos, in addition to religion, prejudice, economic (oil) interests and military powers, are involved.

The Iranians are manipulating the UN and other Western bodies involved in trying to resolve the conflict. Each side is rattling its sabres. The Iranians will probably persist until the last moment before agreeing to a certain compromise, in order not to lose face.

On the other hand, strong pressure by the USA and Europe that causes the people of Iran to rebel and replace their regime with a more liberal one may resolve the problem.

Multi-front negotiation tactics are applicable in this case. Such tactics may include direct and indirect approaches and interventions involving both the governing regime and the people of Iran. The CIA is secretly funding militant ethnic groups in Iran to put pressure on the regime. Simultaneously, economic and other embargos are being enforced.

History has taught us that in most cases of severe conflict between countries, where so many areas of disagreement and hostility are involved, only war can end the dispute.

Is war inevitable? Will this complex dispute end with a compromise? Time will tell.

In summary, negotiation is a dialogue between two or a group of people.

The main intention is to reach an agreement and understanding or to resolve conflicts between the parties. It is essential that the parties enter into the negotiation phase willingly and that they commit to accept and execute the resulting outcome.

Normally, a good negotiation process is terminated by a compromise acceptable to both sides. The definition of a good negotiating process, however, is not when one side wins but when all negotiating parties come out as winners.

This can happen only when a win-win situation is achieved.

What does it take to persuade people? How does a lawyer cause an arbitrator, judge, businessperson or other lawyer to reach the desired conclusion of their own volition?

> *If you wish to persuade me, think my thoughts, feel my feelings and speak my words* (Cicero).

9. Scientific and Philosophical Insights

In this chapter certain scenarios, events and philosophical insights are described from the point of view of the author.

9.1 Spiritual evolutions and the power of belief

Spiritual evolution is the philosophical, theological and spiritual idea that nature and human beings evolve according to a predetermined cosmological pattern or potential.

This *high* spiritual evolution, unlike the deterministic biological evolution of physical form, is surrounded by diverse theories. They attempt to describe our existence and the development of the individual. They view high realities as ideal, mental or spiritual and some claim that mental and physical reality are the same.

All these theories can be seen as teleological. Teleology is a philosophical account which states that final causes exist in nature. We say that a process or action is teleological when it is committed for a final cause. Teleology was explored by Plato and Aristotle, and later by Immanuel Kant (1724–1804) in his *Critique of Judgment*. Aristotle argued that animals exist for humans' use.

We demonstrate *extrinsic finality* when we seek something external to ourselves, such as the happiness of our spouse. If the external thing had not existed, that action would not show finality.

On the other hand, an action has an *intrinsic finality* when it is performed for the sake of something not external to itself: for example, if we want to be happy for the sake of being happy. Many philosophers, scientists, and educators have proposed theories of spiritual evolution.

The spiritual area sees a lot of demand, which is partly satisfied by mystic *clinics* and *care providers*, specialized in chorology, numerology, astrology, tarot, coffee, neuro-linguistic programming, the emotional freedom technique, astro-psychology, and other conventional and unconventional healing methods for the soul and body.

The power of belief - belief is a basic human right. We humans have a need to believe. Sometimes, that need has a power of its own and we follow it without the ability to stop, avoid it or ignore it.

When we talk about belief most of us think immediately about religion and faith, but belief is much more than that. In fact, faith is merely a subset forming part of a wider concept of belief.

It is true that religion is the common concept relating to belief and without it religion would not exist, but belief is a powerful energy that drives us to perform or to avoid many activities.

Belief cannot dwell in a realistic realm. Where there are facts and evidence, belief has no true basis for existence.

Faith and belief may exist in a world where reality, facts and evidence are replaced by personal emotions and irrational views or observations of things.

Sometimes factual events or archaeological findings or other material proofs can be ignored by a blind and fanatical believer. Such a person is unwilling to consider the possibility that their and their ancestors' belief may be irrational or based on an imaginative or speculative scenario. Faith relies upon its believers. Believers in a common faith are united under its umbrella.

Faiths have historically used force to persuade and make people believe, in order to increase their membership.

Religion and science have a 'love-hate' relationship. Science is based and relies on proofs, reason, empirical evidence and rationalism. Religion is based on faith and sometimes blind belief in a non-real or non-existing entity, belief in revelation and sacredness. For the believer, however, that entity may exist and be very real.

We could say that both religion and science are pursuing and seeking knowledge, albeit from a different perspective.

Human history and philosophy suggest, however, that science and religion are more in conflict than in agreement.

The search for a link or correlation between spirituality and neural phenomena is known as spiritual neuroscience or neuro-theology. Spiritual or religious experiences can be explained by neuro-theology and scientific neurological brain research findings. Certain religious activities or beliefs can be associated with temporal lobe epilepsy seizures, during which an enhanced emotional response to religious words may be observed. This finding indicates that the medial temporal lobe might be involved in generating some of the emotional reactions associated with religious stimulation.

Belief is fostered in many ways. Faith-related belief is usually maintained for a long period, or even a lifetime, if taught in early childhood.

Belief can be adopted if a person is influenced by a charismatic 'guru' or preacher.

Other types of belief may depend on the local community, tribe or country that one belongs to.

We may adopt or change our beliefs when we are exposed to certain emotions, such as sexual implicated advertisements.

The power of advertisements, particularly those with sexual implications, can lead us to adopt new beliefs and habits, whether as consumers or as believers (in certain events, or in certain people, such as politicians...). We may be influenced by certain traumas, especially brain-related injury or disease.

Belief is originated and 'cooked' in our brain. Our thoughts are very powerful, whether positive or negative. Accordingly, the power of belief has a tremendous influence on our life.

Recognition of this fact should mean we try to control and master our thoughts. The problem is that when emotions are involved negative thoughts and beliefs are all too common.

Can we modify and change our beliefs?

The answer to that question is yes.

Belief from a historical perspective is a human evolutionary process. The more knowledge and experience we accumulate, the more tools we have and the greater our capabilities to change and modify our beliefs.

There are many ways to effect change but we have to assist in the process by doing things in a different manner.

Each person can develop a method in their own way. If we realize that our belief is limiting us or is a barrier to certain goals or achievements, we have to try to remove those obstacles by acting without hesitation or fear of failure. If we want something or want to become a different person, we should act or simulate the scenario as if we have that thing or as if we are already that person we want to be.

If we do not believe that we can do it, then probably we will fail. We have to overcome the fear of failure, which is the limiting factor.

We will be able to control the power of belief by training ourselves to think positively, acting and behaving as though we have already achieved our goals. We have to believe in our ability to make that change.

The power of belief states that: 'We can be whatever we *want* to be or whatever we *believe* we can be'.

Belief in clinical environment has a substantial impact on ill and injured patients.

There are many cases of 'miraculous' recovery after cancer or crippling accident.

Biofeedback as used in neuroscience has shown cases where changing the patient's belief relieved or even cured the diagnosed complaint. This is a remarkable situation, whereby the patient may cure him/herself, if s/he is strong enough to believe that it is possible.

Placebo effect is another clinical area where the power of belief is obvious and has been proven to have a real impact on patient recovery. Patients are given a 'real' drug or a 'false' (placebo) drug treatment. The placebo phenomenon is that they do not know this and accordingly *believe* that they are getting a real and effective drug, which will improve or cure their illness. Thus believing is healing, the true power of belief.

From this phenomenon it follows that if we are cured by a fake drug then our belief could be applied successfully in other areas of our lives.

If we, or more precisely our brain, can be tricked like this, then we should apply the trick to boost self-esteem and belief in ourselves and in our abilities to overcome the many obstacles that life presents us with. Chapter 3 on love presents a similar method of belief: 'Fake it till you make it'.

Our immune system is built to protect us from many external threats; however it fails when emotion and stress are involved.

If we permit 'bad energy' to enter our brain, we are flooded by bad emotions that can affect our physical body. This is the negative power of belief.

The link of mind-spirit and body is obvious. We should realize that the power of belief can also work against us and may be a destructive force. Therefore, we should avoid negativism and seek positivism. Above all, we have to believe in ourselves.

Willpower is one of the significant human driving forces. We know that there is a power in our will, as in the saying: 'Where there is a will, there is a way'.

This willpower helps us to overcome the many difficulties and obstacles in our life. It is one of the major components needed for success.

The question is whether willpower is the same as free will.

Free will can be defined as the act of choosing without any constraints. By choosing, I mean acting or thinking freely and independently.

Constraints can be classified into several groups of influence: psychological, social, physical, medical, ethical or other influences that might affect our will.

In addition, our will may be linked to and dependent upon other factors such as religion, science or ethics. In neuroscience, for example, free will can help to explain human behaviour.

The existence of free will, however, is debated by philosophers rather than scientists.

In studying the brain, we can see in real time under MRI (magnetic resonance imaging), the process of decision-making. When we wish to move our hand, the brain sends the appropriate signal.

The question is this: 'Is there a time gap between the instant the decision was made and the time when the actual signal was sent?'.

Recent studies show that there is a time that could be as much as half a second or more. This finding shows that the brain makes a decision some time before we are aware of it, implying that we do not have free will. This scientific finding may force us to redefine the concept of free will.

9.2 Time relativity of evolution

If we examine the evolutionary processes in the history of humanity, we may notice an unusual development whereby the 'amount' or 'value' of the development is linked to chronological time.

We can categorize the rise of technological innovations into two major time intervals: from creation, and for the purposes of our discussion it is immaterial whether it was five thousand or fifty million years ago, until 1900, and from 1900 until today.

When we examine historical, economic, scientific, technological, artistic and cultural achievements, we see unusual differences between the two time periods.

The differences between the two time periods are palpable, especially if we consider the insignificance of 200 years compared with the time that has elapsed since creation.

The technological advances described earlier will generate polarization in our life, extreme differences, especially in the human social area.

Whereas the insignificance is in terms of time units passed, there is enormous significance in terms of evolution and actual scientific progress.

If we further divide the second time period and observe the cumulative evolution post WWII, we see remarkable technological evolution.

The breakthroughs are particularly apparent in computer technology, but also in biomedicine and obviously in communication, which is based on computer technology.

This exponential evolution in science can be attributed to individuals, like Leonardo, Newton, Einstein, Teller, Von Neumann and others. Progress continues apace.

Futurists predict even more significant advances in all aspects of our life. Futurists or futurologists are scientists and other academic and non-academic researchers who predict life in the future, e.g. the shape of human society.

The global trend and other areas, such as risk management and new emerging markets, are predicted by these individuals, who work in interdisciplinary areas. In the nineteenth century, futurism had a theological context and there was a certain resistance to change, and suspicion and even rejection of new technologies.

The author believes that the rapid technology evolution will continue in future. Progress will be significant in computers, telecommunication and biomedical areas.

At the same time, however, the polarization mentioned earlier will widen and may go to extreme differences, especially in the human social area. Unfortunately, all those technologies that led to the progress and changes mentioned above are not shared proportionally in the evolution of homo sapiens.

The fact that the wealth of knowledge and progress is not distributed as equally as it should be has had and will have a significant impact on our life.

This will be evidenced by the rise of primitive non-democratic regimes, in fanaticism and in extreme religious terrorism. Over-populated third-world poor and primitive countries will pose an even greater problem.

The author predicts that we will face a new wave of radical worldwide Islamization, especially in the Middle East.

Although an increasing number of Muslims want a modified democracy to accompany Islam and Sharia, the religious law of Islam, there are many who cannot and will not accept democracy as their way of life.

Islamic opposition to Western democracy, such as the Muslim Brotherhood, will pose a significant threat to world peace, especially in the Middle East.

9.3 Islamic invasion

The worldwide Islamic invasion was envisioned by Nostradamus in the mid-1500s. Among other prophesies, he described Anti-Christian Muslim forces in Iraq and Syria, and persecution in the Muslim countries of Asia, especially Turkey. From Israel the war extends to Western Europe and there will be a Third World War using nuclear missiles...

In his apocalyptic prophesies, Nostradamus described the Islamic demographic movement or invasion into Europe. This prophesies has become a reality.

From 1990 to 2010 there was significant demographic change in most European countries. The natural immigration from Islamic countries into Europe has significantly increased the size of the Muslim minorities in those countries.

Muslims struggle to convert the world to Islam. In the past, Christianity and Islam wanted to spread the word to everybody everywhere, by force if necessary. Today Christians and certain other religions no longer wish to use force in the name of their faith; however, Muslims do.

In this age of political correctness in Europe and in the USA, this invasion is generally ignored in the name of multiculturalism. Muslims even have right-wing acceptance and support from anti-Jewish elements in Europe.

Time is working in favour of the Muslim minorities. As Europe becomes older, the Muslim communities will develop into the majority as their demographic increase is about three times greater than that of indigenous Europeans. Yet another prophesy of Nostradamus is being fulfilled.

There are those who do not believe in prophesies in general and particularly not those written in riddles, which may have multiple interpretations.

When one considers the many prophesies that have proven correct, however, it is rather difficult to dismiss them as nonsense.

La grande bande et secte crucigere,
Se dressera en Mésopotamie:
Du proche fleuve compagnie lege,
Que telle loy tiendra pour ennemie. [CIII, Q61]

The great band and anti-Christian sect of Muslims will rise up in Iraq and Syria near the Euphrates with a tank force and will hold the [Christian] law to be its enemy.

This particular interpretation was by Jean-Charles de Fontbrune, translated by Alexis Lykiard, printed back in 1984. A later interpretation is:

The great host and sect of the crusaders,
Will be massed in Mesopotamia:
Of the nearby river the fast company,
That such law will hold for the enemy.

This Islamic evolution seems to be inevitable and at the same time we can see the long-term influence of technology on demographic changes and the desire of people for freedom.

This evolution is particularly noticeable in countries where people are oppressed by civilian or military dictatorships.

9.4 Diversification of democracy

Owing to the expected demographic and geopolitical changes, we can see clear preferences for democratic rule. There are many types of democracies. The following is a partial list:

Constitutional democracy - a democracy governed by a constitution.

Defensive democracy - a democracy where there is a need to limit certain rights, even freedom, in order to protect the institutions of the democracy.

Religious democracy - a democracy where religion is in the centre of the public arena.

Parliamentary democracy - a democracy where the cabinet, which is the executive branch of a parliamentary government, is headed by a prime minister.

Multiparty democracy - multiple political parties may gain control, usually in a coalition.

A dictator is a ruler whose powers are unconstrained by external or superior law. The dictator is the law, so he can take whatever actions he wants, even if they are considered to be illegal.

Can two different ruling systems be combined? The answer is yes; it is a very interesting type of democracy called a *democratic dictatorship*.

Democratic elections place a dictator in a position of power. After he assumes office, the dictator will use the extensive powers that he is able to exercise. It is entirely possible to have a democratically elected dictator. He will be duly elected to office and will be able to exercise dictatorial powers.

Mao Zedong argued that democratic and socialist revolutions could be united in a single phase and did not necessarily, as many claimed, require two separate consecutive revolutionary phases. He named this single phase the New Democracy.

The intention of Mao's New Democracy was to get rid of feudalism and strive for national independence from colonialism.

Following the establishment of the New Democracy according to Mao, the ideology of China became socialist and later communist. Actually, this revolution was constructed of three and not two phases. The New Democracy was established in phase one, followed by socialism and finally communism.

This plan led to the creation of a communist society, one without a state, with no class system and no personal individual wealth.

The New Democracy of China is in fact a people's democratic dictatorship, which can be viewed as a unique and degenerate form of democracy.

The optimal democracy has an elected strong and decisive leader who can take risks for the benefit of the country and not for the benefit of the party.

9.5 The desire to live

A common phenomenon is that when one half of a couple that has been together for a long time dies, the other also dies shortly afterwards

The usual, romantic, explanation is that they loved each other so much that they could not live without each other.

He or she died from a 'broken heart', 'could not live alone', 'was dependent on their spouse'.

Is there a correlation between the death of one spouse and the subsequent death of the other shortly afterwards?

The hypothesis is that the desire to live may have a certain effect on the immune system. The brain that is in control of our body may have a shut-down mechanism, which is activated in certain cases. Those cases are similar to fatal accidents or certain illnesses, where the brain knows that it will not be able to cope. This mechanism may control 'suicide cells'.

In recent years, suicide cells or what scientists define as programmed cell death (PCD) has formed the basis for ongoing biogenetic research. PCD is the death of a cell which is mediated by an intracellular programme.

There are three major types of PCDs. Type I cell death is called apoptosis. Type II is autophagic and Type III is necrotic cell-death.

Cells can be killed by injurious agents or be instructed to commit suicide. If there is a threat to the integrity of an organism by certain cells, PCD is needed to destroy those cells.

Typical examples of such cases are: cells that are infected by viruses, DNA damage, cells of the immune system and cancer cells.

In certain types of cancer cells apoptosis is triggered by radiation or chemicals used for therapy.

What makes a cell decide to commit suicide?

The author believes that it is the imbalance between positive and negative signals sent by the brain.

If there is a lack of the positive signals (no desire to live) needed for survival and/or negative signals are sent meaning 'no desire to continue to live', the shut-down mechanism may be activated.

There have been numerous reports of cases where patients recovered miraculously after clearly being diagnosed with cancer. This phenomenon may be explained by the activation of the PCD mechanism by 'desire to live' positive signals.

In some cases, viruses that are associated with cancers may use tricks, like producing a protein that inactivates the apoptosis signal. In such cases the cancer cells will not only continue to live and proliferate, but they will become more resistant to apoptosis. Further understanding of those tricks and decoy molecules generated to protect cancer cells would enable researchers to reactivate and overcome those protective tricks in order to destroy dangerous cancer cells.

The author also believes that future research on destroying and removing cancerous cells might be implemented in two phases. The first phase would be to distinguish, mark and identify cancerous cells. In phase two the target would be to activate suicide cells in the selected area or group of cells and bypass the existing protection of the cancer cells.

Strong psychological and family support is essential for such a recovery. In addition, the patient must believe in and hope for a healthy and bright future.

9.6 Virus of the mind

Virus of the mind or *brain virus* can be defined as the method or algorithm that enables ideas, messages or trends to be implanted in the brain.

It is done in such a way that people genuinely believe that it is their original thought or will. The brain can be stimulated by many external media, such as advertisements to purchase goods.

Brain virus can affect the way we live or even induce suicide. Religious beliefs are carefully selected brain viruses.

Throughout history people have committed atrocities in the name of their God.

Strong and weak people infected by the brain virus will react completely differently. Brain virus may use threats, fear, pleasure and other senses and desires in order to achieve and act upon the selected target.

Anti-brain-virus (ABV) is the personal ability to ignore and overcome implanted viruses. Phase I - implant, Phase II - evolve, Phase III - act.

ABV will go into action after it recognizes that a message on which we are about to act is a virus. The defence mechanisms against mind viruses can assist us in changing or eliminating those thoughts which are imposed upon us by the implanted viruses.

Certain memories fade faster than others, possibly depending on our need or willingness to remember or forget.

In early experiments on rats, scientists found a certain enzyme which caused memory to fade more rapidly than usual. This finding suggests the possibility of selectively erasing certain undesired memories. Does it work the other way, too? Can we boost or enhance our memory?

Future research could lead to genetically engineered brain virus, which will enable such memory enhancement.

A spin-off of brain virus is brainwashing, which originated during the Korean War in the 1950s.

Brainwashing is defined as the application of certain techniques, usually coercive, to manipulate or change a person's beliefs, values, perceptions and judgments. Subsequently, it affects the behaviours of that person.

The purpose of brainwashing is usually to change someone's political, personal, or religious opinions and beliefs.

The Chinese used brainwashing methodologies under the Maoist regime. Their aim was to transform individuals with a reactionary imperialist attitude into 'right-thinking' members of the new Chinese social system.

It was largely done in two phases. First, the person's psychic integrity was broken down by dehumanization, and then when they were 'clean' they would go through certain rites.

Cults in general use brainwashing techniques to recruit members into their organization. The 1960s to the 1970s was a decade when mind control theories were applied by religious movements, especially in the USA.

The success of many a totalitarian regime can be explained by the use of brainwashing and mind-controlling techniques.

Brainwashing or mind control theories are researched from neuroscience and social psychology perspectives. The activation of brainwashing can be done by manipulating the prefrontal cortex of the brain.

Project MKULTRA was the code name for a covert, illegal CIA human research programme run by the CIA's Office of Scientific Intelligence during the 1950s and 1960s.

This project used scientific, medical and psychological methodologies to manipulate individuals' mental states and alter their brain functions.

Diverse and complex combinations of chemicals, drugs, hypnosis, isolation and sexual abuse were used.

Here are selected quotes from the New York Times.
(August 2nd, 1977)

> *The C.I.A. leaders were certain the Communists had embarked on a campaign to control men's minds and they were determined to find a defense, setting out in earnest the next year-1950-with Project Bluebird, which evolved into Project Artichoke, then became MK-ULTRA, MK-DELTA.*
>
> *With each code name change, they broadened their sweep, until there remained virtually no avenue of human behaviour control they were not exploring.*
>
> *There was an "urgent need," the C.I.A. and other intelligence agencies argued, to develop "effective and practical techniques" to "render an individual subservient to an imposed will or control."*
>
> *One of their longest running goals was to develop a way to induce amnesia.*
>
> *They wanted to be able to interrogate enemy espionage agents in such a way that neither the agents nor their superiors would know they had been compromised, and they wanted to be able to wipe clean the memories of their own agents after certain missions and, especially, when they were going into retirement.*
>
> *They were interested in simple destruction, too. As with the other business that made amnesia so attractive, they wanted to be able to get away with murder without leaving a trace.*

The following CIA document was declassified and released to the public on December 1994:

PROPOSAL

It is proposed to continue research on problems critical to a clarification of the fundamental aspects of the stimulus-response relationship in biological systems.

Studies will be conducted utilizing chronic animal preparations with permanent septal electrodes.

Investigations will be conducted utilizing chronic electrode preparations coupled with selected classical types of conditioning.

Studies will be conducted on the geography of the brain in selected species of animals to determine the locus in which stimulations will produce specific reactions.

A careful literature survey, especially of the Russian literature of foreign research in this area will be conducted.

The evolution of human communication – humans cannot exist without communicating with each other for various purposes.

Different forms of communication are required for everyday life, entertainment and many other areas.

Communication can be verbal, telepathic, musical, use signs or SMS, electronic and other non-electronic means, and use visual and audio signals.

In the prehistoric era, fire, smoke signals, beacons, drums and horns were used.

The mail was introduced about the sixth century BC and pigeons were used to send post from the fifth century BC.

Semaphore, which is a system for communicating by means of visual signals, using towers with blades, was introduced in the fourth century BC and the heliograph, which is a wireless solar telegraph that uses signals based on Morse code and mirrors reflecting flashes of sunlight, was used from 490 BC.

The maritime flag system was introduced about the fifteenth century and signal lamps were used from the nineteenth century, mainly by the navy using Morse code.

The electrical telegraph was in use from 1836 and the telephone, photophone (radio-phone) and the radio were invented in between 1876 and 1896.

The television has been in use since the late 1920s, the videophone from the 1930s, telecommunication using fibre optics since 1964 and computer networking since 1969.

Cellular mobile phones came onto the market in 1981 followed by email and the Internet, and satellite phones have been in use since 1998.

What next?

It is believed that, as the technology improves, these communication devices may become part of our body, rather than an accessory which we carry or wear.

We could become wired organisms and meet each other in a new medium in cyberspace.

9.7 The polarity of everything

The law of polarity states that everything in life has its own polar opposite. The theory of relativity does not distinguish between good and bad or large and small.

The law of polarity claims that they are the same but on opposite sides of the spectrum and they are relative to our perception. Our perception of polarity is much more complex, however.

We would have great difficulty recognizing good without bad, small without large.

All these opposites require relative comparison so they can be perceived properly. Everything in life has a corresponding opposite. Our experience and perception tell us what side of the pole we see or want to be.

Moreover, pairs, couples of polarities, cannot exist in a singular form as we would not be able to experience or see the other.

If we did not experience sadness we would not know how to experience joy. There is no pleasure without pain.

Failure and success are the same thing at opposite ends of the spectrum. Within failure there is the potential for success and vice versa. We should not resist failure, as we will get more of it. Accepting and understanding failure opens the door to its opposite, namely, success.

Other examples of polar opposites are: night and day, sad and happy, pain and pleasure, failure and success, black and white, love and hate, yin and yang, male and female.

9.8 Life-logging

Life-logging is the recording of every single minute of our life. It is an audio-visual recording of the way we see the world around us throughout our entire life.

The full life log is a document containing not only the above-mentioned large audio-visual data base, but all the videos, pictures, documents and mail, sent and received during our lifetime.

Digital storage technologies are available at relatively low cost, of high quality and have adequate resolution.

The basic digital database for 100 years' continuous recording comprises 100y*365days*24hours*60minutes*60seconds of data with the desired resolution.

This corresponds to only 3,153,600,000 bytes or resolution units.

In future, this stored information could also include our location via the GPS system, physiological data and even our feelings and senses such as smell.

All the recording media required to perform the logging, such as the camera and the sensors, will be wireless and implanted. We could become a kind of human cyborg.

9.9 Experience transfer to the next generation

When Einstein died, his brain was preserved and studied. The irregularities of his brain structure compared with the average brain were noted.

Will evolution and progress enable us in future to transfer our unique capabilities to the next generation?

Actually, evolution is the passing on of certain genes from one generation to the next through natural selection.

There are two different ways in which genes are transferred according to the rules of evolution.

One is the passing of genes directly from parent to offspring.

The other is a laboratory product of bacteria and other species, which are not capable of sexual reproduction. They evolve by sharing selected genes via direct cell to cell contact.

Will we be able to transfer our wisdom, intelligence and even our experience to the next generation?

Mind uploading is the transfer of our mind or brain content to another brain, computer or robot and it is a popular theme in science fiction. Humans generate copies of themselves by using a human-like android body and uploading their mind just as we generate backups of our important data.

This fictional mind upload generates ethical, legal, philosophical and identity problems.

In addition, there could be more problems in terms of the financial and computing implications of maintaining the mind in the transferred brain or backup computer

With mind uploading, when it will be available, humans may achieve immortality.

9.10 Neurophilosophy

Neurophilosophy is a relatively new scientific field in neurosciences and philosophy. Combining these two exciting research areas using interdisciplinary methods and tools may lead to new and exciting findings.

Using empirical methodologies from neurosciences, we may find solutions to philosophical problems and by using methods of philosophy of science we may find ways to clarify neuroscientific results.

If we want to explain our mental activity and are not explicitly interested in explaining our brain function and structure, we look to the theories of neurophilosophy.

Neurophilosophy and education are an interesting and a promising area of research in terms of the interaction and relationship between educators and students.

On the one hand, we have the educator-teacher, who has knowledge and the experience on certain subjects and may be perceived as the 'transmitter'.

On the other, we have the student, who has the need to be educated. S/he needs or wishes to acquire knowledge for diverse purposes, such as an academic degree for professional use. The student may be perceived as the 'receiver'.

Actually, the process of learning-studying is a neuro-interaction between two brains, an indirect transmission using language as the medium of communication. This process uses brain-language-sound generated on the transmitter side and brain-language-ears on the receiver side.

The receiver needs to translate, process and store the message in his/her brain. As this process is indirect between the two brains and involves several elements and organs, it is exposed to 'noise' and interference. One such example is ADHD (attention deficit hyperactivity disorder).

There are other interfering elements in this process and the ability to receive the same transmission in a class environment varies from student to student.

Willingness to learn is a major factor in the student's ability to acquire the desired knowledge transmitted by the teacher. The student needs to open his/her receptors and receive the information in a certain form commensurate with his/her ability to understand.

The teacher must be in a transmitting mode and at the same time the student must be in a receiving mode. If the student is not in the receiving mode, the teacher's responsibility is to generate the appropriate conditions so that his/her messages will be properly received, at least by most students.

This type of education is more student-oriented, unlike the standard 'fit-all' methods.

If the receiver-student has receiving problems, the transmission and the transmitter must be checked. The ability to receive an unclear or encoded transmission depends on the receiving brain and its ability to decipher, analyse and store the message. It is, however, primarily dependent on the quality, clarity and authenticity of the message transmitted by the teacher.

Understanding occurs when the receiver signals the transmitter, 'I have got it'. This acknowledgement can take many forms, such as a written or oral test, when the teacher can assess the student's understanding.

So how are those messages stored by the brain? Are all those messages stored in one location in our memory brain?

We do not yet know the answer but we know from neuro-imaging experiments that this process is neuron-dependent. We can observe 'action' or 'light-up' of our synapses during the learning process. This fact suggests that information is not stored at one specific location but spread over a wider area of neurons.

Another supporting fact is the ability of a patient suffering from post-traumatic brain injury or loss of certain parts of the brain to recover. If that information was stored in one location only, such recovery would be impossible.

In addition, we know that our memory is associative, which means that we can recall certain scenarios or information from our 'experience' data base memory by using related and unrelated triggers, such as sound, smell, touch or other visual or audio experiences. This proves that there are many interconnections in the same cluster of data to be retrieved.

The recall process enhances and strengthens the neuro-connections between the relevant cells, which accordingly will keep the data stored viable, accessible and fresh. Learning will generate similar enrichment of the neurons and their interconnections.

Our brain consists of trillions of neurons, with a huge number of complex interconnections. What differ from brain-to-brain are the types of neurons and the specific neurochemical interaction among the neurons. It is interesting to note that the structure of clusters of neurons and their specific interconnections may have an effect on one's ability to learn and an influence on speed of understanding and reaction time to intellectual stimulations.

At birth, our brain is very plastic, that is, its capability to process and store sensory information is very high. Neuronal connections are generated, broken and regenerated, which suggests that early educational and environmental stimulations are essential for the child's evolution. This is the critical period of the development of the child's linguistic, cognitive and social abilities.

A classical question is whether the infant brain is empty, a tabula rasa, at birth. The Greek philosopher Aristotle (fourth century B.C.E.) was probably the first to introduce the tabula rasa (blank slate) idea.

According to the tabula rasa theory, an infant's brain is empty of mental content, which will be acquired later with experience and perception.

Although the 'tools' or the brain cells are already formed at birth, only after gaining experience will we see the generation of neurons' inter-connections.

As Aristotle and subsequent supporters of his theory were not privy to recent genetic discoveries, the tabula rasa theory may not be applicable or accepted as a deterministic valid concept.

Today it is believed that a child's cerebral cortex is pre-programmed to enable the processing of sensory input, emotions and environmental stimulations.

The author does not support the tabula-rasa theory and he believes that there are genetically transferred data or imprints. Those genetic imprints may have a clear impact and influence on the child's behaviour and even on its brain's ability to process and store information.

Technological evolution is still happening in computer technology, when we communicate with computers not only via a keyboard, mouse or touch-screen, but also via sound or voice. Similar evolution is happening in medicine such as brain-activated devices based on EEG (electro-encephalo-graphs).

In future we could experience direct brain-to-brain transmission similar to telepathy. Telepathy derives from the Greek ('distant experience') and it is a kind of mental transfer from one brain to another. As it is not a clearly reproducible phenomenon, the scientific community has not reached consensus. Telepathy is well accepted, however, albeit largely used in science fiction.

As many science fiction scenarios became reality in time, however, the author believes that some brain-to-brain communication will be possible in future. Neuro-imaging is one of the scientific areas where this type of communication is being researched and interesting results are anticipated.

The conclusion is that educational methods must correspond and comply with our brain function and its ability to store information and not on a dogmatic rigorous unified system as exemplified in most schools.

9.11 Brain-to-brain communication

There are many brain research centres worldwide. Advances, innovations and breakthroughs are published in several scientific journals and brought to the popular audience in style that all can understand.

From time to time we see remarkable progress in the understanding of brain functions. Unfortunately, despite this progress we are still far from deciphering our brain structure and its complex, diversified and mysterious functions.

Among the many neuro-scientific studies is interesting and promising research on basic communication methods and deciphering the way in which the brain encodes and decodes spoken language.

Human communication skills and ability have evolved in a unique and miraculous manner.

The lines written on this page are transferred from the author's brain to the reader's brain, via written characters, forming understandable formal textual words and sentences in a common language.

In this brain-to-brain transfer, no acoustic input is involved, unless the text is read aloud.

Researching the brain activity when acoustic input is involved is a very complex procedure. Aside from the scientific problems, we face a clinical and technical problem, as we are unable to experiment on live human brains.

Non-invasive methods are insufficient to accumulate the data needed for such a study.

Recently, a unique opportunity was given to brain researchers to obtain experimental data directly from the human brain. Researchers observed neurosurgeons while they were performing routine neurosurgical procedures to treat epilepsy patients.

During this neurosurgical procedure, the brain is exposed in areas such as the cortical surface, which allows direct measurement of the neural activities.

This unique opportunity permits the data acquisition required to perform brain studies, such as speech recognition.

Preliminary results were intriguing. From the neural signals accumulated, it was possible to reconstruct certain words. These findings indicate the possibility of reading acoustic speech directly from the brain.

It should be noted that the neural signals acquired were two-dimensional surface electrodes.

The data are much more accurate and relevant than the data received from external surface electrodes on the scalp. The ultimate aim would be to collect three-dimensional deep neural signals.

The neurological community recognizes that signal messages sent by healthy nerve cells/neurons are different from those sent by damaged, inactive, sick nerve cells.

As the local EEG signal is lower (weaker), it is absorbed by its neighbouring regions and they are not necessarily reflected in the cumulative EEG detected at scalp level by electrodes. In addition, local regional nerve cells, which are inactive, damaged or suspect, send certain messages which may correspond to early warning.

These local messages are lost in the cerebral activity signals. They might be detected at a much later stage, when more damaged cells join the region, or when symptomatic effects take place, that is, when it is too late.

The challenge is to separate and amplify these local messages and to generate a new imaging diagnostic tool for brain mapping of regional cerebral activity (RCA).

This RCA signal in addition to standard EEG and other neuro-clinical findings may enable more accurate localization of inactive or damaged nerve cells and consequent early detection of brain abnormalities.

This new type of algorithm, in addition to the specific function as described above, may constitute the basis for the generation of the three-dimensional neural signals needed to read acoustic speech and other data directly from the brain.

10. Chronic Pain Detection

The American Academy of Pain Medicine defines pain as: *An unpleasant sensation and emotional response to that sensation.* The web version of the Encyclopedia Britannica defines pain as:

> *A complex experience consisting of a physiological (bodily) response to a noxious stimulus, followed by an affective (emotional) response to that event. Pain is a warning mechanism that helps to protect an organism by influencing it to withdraw from harmful stimuli. It is primarily associated with injury or the threat of injury, to bodily tissues.*

Pain is an individual sensation that can best be described or defined by the person experiencing it. It may cause distress and discomfort and it is usually described as aching, pinching, throbbing or stabbing.

We can distinguish between two basic types of pain - *acute* and *chronic*.

Acute pain lasts a relatively short time. It is a signal that indicates when a body tissue is being injured. The pain generally disappears when the injury heals.

Acute pain results from disease, inflammation or tissue injury. It may appear suddenly, such as after surgery trauma, and may be accompanied by emotional or other distress.

The cause of acute pain can usually be diagnosed and treated. In certain cases it may become chronic.

Chronic pain ranges from mild to severe and lasts for long periods of time, usually more than three months.
Chronic pain is associated with the disease itself. It may be aggravated by psychological or environmental factors.
The cause of chronic pain is not always evident. In certain cases, it is associated with chronic conditions such as arthritis, fibromyalgia or lupus with symptoms such as swollen joints, unexplained fever, extreme fatigue, sleep problems or red skin rash. Chronic pain syndromes, in particular, are complex and their effective treatment will often involve coordinated, multidisciplinary consultation.

In contrast to acute pain, chronic pain can be mysterious, intractable and often very expensive to treat. The complexity of chronic pain stems from the fact that it is a biopsychosocial condition, which occurs in various forms.

Since pain is a biopsychosocial condition, all aspects of the condition must be treated. Assuming that a condition is 'all in the patient's head' overlooks the possibility of real pain.

On the other hand, failure to assess the psychosocial factor can also lead to longer recovery. The complex nature of chronic pain disorders makes it impossible for a single professional to treat it successfully.

We can distinguish between *peripheral* and *central* pain. *Peripheral pain* originates in the peripheral nerves or in muscles, usually via trauma.

Central pain arises from pathology or dysfunction of the central nervous system (CNS). It is primarily owed to structural changes in the CNS such as spinal cord injury, multiple sclerosis, stroke and epilepsy.

Inhibition of pain is important and needed especially when our safety is more important, e.g. when we are running away from a dangerous situation.

The purpose of pain is to tell us via our brain when something needs to be done about a damaged area. The brain will tell us to pay attention to the painful area or to ignore it.

The information transmitted by the brain travels to the spinal cord or brainstem via electrical impulses in fibres of spinal or cranial nerves. The signals pass electrically to higher CNS levels.

Therefore, monitoring these signals in real time may help in monitoring, detecting and verifying pain.

Pain and gender. Recent studies using positron emission tomography (PET) brain scans of patients during pain stimuli showed different brain responses between men and women.

Several areas of male and female brains responded differently to the same pain stimuli. Female brains showed more activity in emotion-related centres, whereas males responded in the cognitive or analytical regions.

The differences may relate to our evolution process and the different social tasks of males and females.

Women often have high pain experience levels, but lower pain tolerance. Their sensitivity to pain is affected by many factors such as biological or inherited conditions and hormone levels.

Pain and Animals. The presence of pain is detected by the observation of change from normal behaviour.

Pain may be evident as a limp or a change in gait, withdrawal or protection of an injured part, abnormal postures, licking, rubbing or scratching at an area. Signs of pain and distress particular to rodents include eating too much, chewing toes and feet.

Signs of pain may be subtle such as a change in respiration, reluctance to move, apprehension, sudden aggression, inability to rest or sleep normally, or a worried or anxious expression.

After injury patients are referred for orthopaedic care, where the appropriate treatment may resolve the problem. In certain cases, patients will not get lasting relief mainly because of inaccurate diagnosis.

The author is the founder of Imexco General Ltd. Imexco has been involved in the development of medical diagnostic equipment in neurology and cardiology for almost two decades.

Imexco is the exclusive owner of core technologies were developed by the company, the neurology related Neuro-Core and the cardiology related Cardio-Core.

The Neuro-Core was the basis for the successful development of a new neuro-brain monitor, the Neuritor™.

The Cardio-Core contributed to the Neuritor™ with embedded cardiac application and was the basis for the development of the CardioScope™, which is an ambulatory ECG monitor.

The Neuritor™ received FDA (American Food and Drug Administration) clearance for marketing as an EEG (electro-encephalo-graph) device.

In addition, Imexco has developed proprietary technologies which include a new algorithm for the generation of RCA™ (regional cerebral activity) signals, as described earlier.

Another product that partially utilizes the Neuro-Core is the Chronic Pain Detector (CPD™).

The Neuro-CPD™ system is based on a multi-channel physiological signal processing algorithm in real time and can also operate under ambulatory conditions.

The major components and modules are:

Data acquisition – collecting physiological signals in real time. Receiving oral information from the patient and generating a specific adaptive patient pain profile.

Data processing and analysis – the acquired data are processed and analysed by sophisticated heuristic artificial intelligence based algorithms. All data are correlated and transferred to the diagnostic module.

Diagnostic reporting module – this is the pain-reporting module. Neuro-CPD™ can localize and classify the source and type of pain, psychological or biological, which in certain cases may make surgery unnecessary.

This device can save billions of dollars on inappropriate procedures, assist patients and save insurance companies paying out on unjustified claims. The system can monitor and document the effects of physiotherapy, pharmacological treatments and other interventions for clinical as well as for research purposes.

Neuro-CPD™ is a complementary tool for the physician to diagnose correctly and recommend appropriate treatment. It is not intended to replace the physician nor does it perform any direct clinical assessment. It reports accurately certain vital signs so that in conjunction with the physical and psychological effects and other symptoms of pain the physician can recommend the right treatment.

11. Epilogue

The diverse philosophical and other subjects selected for this book were based on the author's experience and personal interests, as touched on in the prologue.

Each of the subjects was described and developed as far as possible, with due regard to the reader's time and patience. Extensive literature is available for those who wish to deepen their knowledge and interest in the specific subjects raised.

It is acceptable and even predictable that we will raise questions about any subject we encounter in our earthly life. The love of wisdom or philosophy leads us to philosophize about our way of life, the many problems we encounter, our values, our existence and our possible future.

The future is unpredictable; however, in the medical area we might try to second-guess a few things. The application of nano-technology to the monitoring, diagnosis, prevention and treatment of diverse diseases is nano-medicine.

This new and evolving area will revolutionize medical science in the near future.

Nano-medicine applications will address problems in the areas of patient monitoring, pacemakers, biochips, medical sensors, insulin pumps, drug delivery systems and glucose monitoring.

In future, diagnostic nano-machines could monitor the human body internally and in real time. They may be equipped with wireless transmitters and send out warnings when changes in the chemical imbalance are detected. Nano-machines could be planted in the nervous system to monitor brainwave (EEG), and other vital signals and functions.

Another exciting area would be implanted drug delivery systems. At a higher level of the application we may see in future nano-robots acting as mini-surgeons on the level of single cells.

We can only imagine what this technology will do for pain management, monitoring and treatment.

Hoping for a better future

www.ingramcontent.com/pod-product-compliance
Lightning Source LLC
Chambersburg PA
CBHW071625170426
43195CB00038B/2123